RETAINING
SOIL MOISTURE
in the
AMERICAN SOUTHWEST

RETAINING
SOIL MOISTURE
in the
AMERICAN SOUTHWEST

by

Kelly J. Ponte, Ph.D.

SUNSTONE
PRESS
SANTA FE

This book contains information gathered from authentic and highly
regarded sources. A variety of references are listed. Reasonable
efforts have been made to publish reliable data and information,
but the author and the publisher cannot assume responsibility for
the validity of all materials or for the consequences of their use.
Any application of recommendations set forth in the following
pages is at the reader's discretion and sole risk.

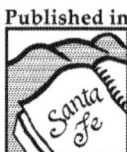

Sunstone books may be purchased for educational, business,
or sales promotional use. For information please write:
Special Markets Department, Sunstone Press,
P.O. Box 2321, Santa Fe, New Mexico 87504-2321.

Library of Congress Cataloging-in-Publication Data

Ponte, Kelly J., 1961-
    Retaining soil moisture in the American southwest /
by Kelly J. Ponte.
        p. cm.
    Includes bibliographical references.
    ISBN 0-86534-411-6 (softcover)
    1. Soil moisture—Southwestern States. I. Title.

S594 .P64 2003
631.4'32'0979—dc22

                                    2003017120

Published in   SUNSTONE PRESS
               POST OFFICE BOX 2321
               SANTA FE, NM 87504-2321 / USA
               (505) 988-4418 / ORDERS ONLY (800) 243-5644
               FAX (505) 988-1025
               WWW.SUNSTONEPRESS.COM

CONTENTS

INTRODUCTION

United States and worldwide populations continue to increase while natural resources such as water continue to decrease. Our waste waters are filtered through the soil and geologic components of the earth where they end up back below ground for our re-use. We use clean, potable water at a rate much faster than the earth's ability to recharge the water supply. Due to our increasing use, depth to water in wells and aquifers is increasing and many wells are drying up. It should be automatic then, to each do our part to conserve, retain, and efficiently use the water resource to ensure a clean, adequate supply for future generations.

States comprising what is referred to as the southwestern region of the United States include primarily New Mexico, Arizona, and Texas. Residents of New Mexico living at elevations of 5,500 to 6,000 feet receive approximately 10 inches of annual precipitation whereas folks living at 6,000 to 7,000 feet can expect about 13 inches. Elevations of 7,000 to 8,000 feet receive about 16 inches, whereas elevations of 8,000 to 9,500 feet receive 22 inches. High elevations of 9,500 to 10,000 feet receive approximately

28 inches of precipitation annually. This is not much moisture when one considers human, other animal, organism, and plant needs.

The fragile environment of the southwest, already at the mercy of an aridic to semi-aridic climate, is often under siege by never-ending drought conditions. Sun intensity and duration, high winds, heat, and prolonged periods without precipitation facilitate erosion and the inability of many plants to thrive. Without adequate moisture, organic matter does not build up and microbes do not flourish. Organic matter, microbial activity, and adequate moisture are key ingredients for a healthy soil. Although most climatic factors that affect the volume of water a soil receives are controlled by mother nature, there are some measures individuals can take to conserve and retain soil moisture. Soil moisture includes precipitation from rainfall, snow and hail, as well as runoff, watering and irrigation waters, streams, lakes, ponds, rivers, gullies, arroyos, and water contained between soil particles . . . basically, any water that comes in contact with soil particles.

The diversity of the southwest, with its variety of geologic formations, landforms and landscapes offers wide ranges in soil moisture content and temperature. Moisture and temperature are key determinants of what grows where. Our ability to grow food is a primary consideration with regards to soil moisture. The food chain depends on an adequate soil moisture supply in order to thrive and reproduce. Humans are part of the food chain.

This book was written in response to the drought conditions in the southwest.

—*Kelly J. Ponte, Ph.D.*
Soil Scientist

1

WHY DO WE CARE ABOUT RETAINING SOIL MOISTURE?

Because we humans depend on plants as a food source, as building materials and supplies, and as a source for many of our medicinal needs. Most plants require moisture mostly in the form of a soil solution in order to survive.

Because we humans would prefer not to be living in a barren, sterile landscape, devoid of plants and surrounded by a cracking, parched soil surface.

Because besides humans, many small living things such as plant roots, rodents, worms, and micro-organisms require soil moisture in order to survive, thrive, and complete their reproductive cycle to ensure survival of their species.

Because many micro-organisms such as bacteria, actinomycetes, fungi, algae, and protozoa depend on moisture in soils. These organisms carry out functions and catalyze chemical reactions that aid in making our world a better place to live.

Because water is a natural resource. All natural resources are precious.

Because the content of water in our soils directly or indirectly affects life.

Because the content of moisture in our soils directly or indirectly affects the soil's ability to filter our polluted and contaminated waters in order to provide all life with potable water.

Because the beauty of the landscape and landforms around us were shaped in part by the effects of water. Large and small scale physical and chemical processes, weathering, and decomposition are affected by moisture.

Because abuse of land and water resources is often the cause of the fall of civilizations.

Because ever-increasing populations place extremely high demands on our natural resources.

Because neither food nor water have substitutes. Our food sources are completely dependent on our water resources.

Because we are using water at a faster rate than we are replacing or recharging our supply.

Because without it, we would die.

2

WATER RELATIONS AND CONSERVATION ISSUES

Water conservation involves the protection, preservation, restoration and management of water resources. Water conservation topics cover many areas including water politics, use and management practices, water quality standards, laws, regulations, and human rights issues. Accessible, usable, clean water comprises less than one half of one percent of planet earth's water supply. The rest is seawater, frozen ice, or water stored in the ground at inaccessible depths. The whereabouts and possession of a land's accessible waters has always been associated with the control of the land and with power. (1)(10)(17)

With the emergence of the Safe Drinking Water Act (1974) and Clean Water Act (1977) the 1970's catalyzed an awareness of environmental issues focusing on water quality, and subsequent remediation and preservation of water and land resources. Recognizing past mistakes were and are crucial to developing workable prevention and remediation techniques to address and manage water issues. Many areas of the country are still plagued by damage caused many decades ago, as remediation techniques take time and money. Currently, we continue to seek balance between water consumption needs of an industrialized nation and exploitation of our water resources.

Water-policy makers and "the water industry" are basically the same entity. The water industry is comprised of private groups and public agencies with interests extending into agriculture, recreation, mining, energy, navigation, and urban development. Although these groups greatly influence water management decisions, individuals see and deal primarily with public agencies to fulfill their domestic and business water needs. Municipal water departments, county representatives, irrigation districts, and even out of state ghost water companies each have leverage, from the levying of taxes and assessment of fees to the ability to make water-related decisions. Engineering, construction, and development interests command investments and investors. Substantial investments are required to develop facilities, build infrastructures, and promote and reinforce interests. And since negotiations require contracts and agreements, the industry is also comprised of water lawyers. Along with engineers, the construction industry, land developers, real estate moguls, irrigators of landholdings, and government and private agency representatives, this diverse group controls and directs water resources and development issues of a given area. They are the water industry. (1)(10)(17)(32)

Many countries experience dire consequences due to the lack of a continuous supply of clean, potable water. Here in the U.S. we are spoiled. Americans are very particular concerning the taste, color, and scent of their drinking water. Today, in an ever-increasing focus on cause and effect, health-related issues and any connection to drinking water, consumers have turned to tap water filtering units and bottled waters, the latter of which is usually treated with granular activated charcoal (GAC) or ozone for taste improvement and visual clarity. (1)(10)(32) Many undeveloped countries grapple with insufficient quantities of clean water to supply their human population for basic needs. These people deal with water borne diseases caused by a contaminated water supply. If they drink from, cook with, wash with, bath in, and defecate in the same water supply, is it no wonder that some of these people face a very different set of consequential living conditions on a daily basis than we do? Here in the U.S., and especially in the southwest, where drought conditions are common, water concerns are not focused on basic human needs such as adequate drinking water. Instead the focus is on whether golf courses should be allowed to water the greens, whether water needs to be supplied in order to save the Silvery Minnow, whether homeowners should or should not be allowed to plant grass on their landscape, and whether or not the watering restrictions provide adequate water for the survival of planted trees, bushes and flowers. This is not to say that these issues are not important issues, just that they are different from those described above. Our priorities, use and consumption habits contribute to breeding contempt towards the U.S. from other nations. Our climate, location, and subsequent ability to irrigate and grow a lot of food creates an oasis compared to many other countries. We are in an industrialized country, who, not long ago and very quickly, went from drawing water from buckets dipped into hand dug wells, to power-drilled deep wells, push button faucets, and pricey bottled water. We've come a long way, but in what direction? Have we taken ten steps forward or ten steps back? Ten steps up or ten steps down? An overview of our direction would appear to be that of a mouse in a maze more than any one vector. This is not meant to diminish our advances and contributions but simply to admit that we've made a lot of mistakes along the way. The U.S uses more than its' share of world resources. If we are unable to discipline ourselves to use less, then the very least we can do is use our resources efficiently and wisely.

There's plenty of water in the U.S., isn't there? Here are a few events to ponder. By the time the Colorado River reaches the sea, it has lost most of its' fluid through its' travels via the U.S. In recent years we have witnessed the event of the cease of flow of the Rio Grande into the Gulf of Mexico. (1)(2)(17) Diversions and outlets created for the purpose of additional sources to supply consumption and irrigation are primarily responsible for the diminished water flow. After World War II, increased practices of putting land into farming and subsequent irrigation depleted the West's largest aquifer, the Ogallala. In some areas, the depletion was of such a magnitude that water users sought to look elsewhere for water sources. Why? Because we divert water, pump it, and use it, just as we do with most of our water bodies. We use fresh water and give the earth back waste waters. We've seen many changes in irrigation practices in agriculture in an effort to minimize evaporation and runoff. This is good. But populations, housing development, and manufacturing continue to increase, which translates to more water use. And our fresh water supplies continue to dwindle.

How much waste water are we putting back into the earth? The manufacturing of most of our products and goods use fresh water and return waste waters. We have lots of goods and services and power at our fingertips. As a country on the whole, we are very affluent. But the price we pay is enormous. The pen I write with was manufactured. As you look around a room, what can you see that was not manufactured? The manufacturing process utilizes raw materials and in return produces products, some necessary products and many unnecessary products. Most manufacturing processes also produce waste water that contains chemicals and substances. In the past, waste water was frequently dumped in the nearest river. Due to public outcry and subsequent watchdog groups and regulations, manufacturers cannot flagrantly discard waste waters. Many manufacturers have turned to removing, filtering, and treating these fluids for re-use. Some paper-making companies, for example, participate in this recycling practice. And I'm sure many do not.

Manufacturing exists due to supply and demand. If we did not use products, there would be no profit in or incentive to produce products. It takes over 100,000 gallons (U.S.) of water to manufacture a car. The computer industry uses over 350 billion gallons (U.S.) of water and produces over 75 billion gallons (U.S.) of waste water per year. (1)(17)(32) I drive cars and use computers. These are items I would prefer not to be without. In our fast-paced, technology driven country I consider them to be necessities. But there is no reason why we can't make them better. More efficiency at the source then all the way down the line. Less waste. More quality and durability. More focus on knowing exactly how a product will be recycled BEFORE it is produced. Every component. All manufacturing waste-waters. Recycling facilities should be commonplace. If it can't be easily and cost-effectively recycled, it shouldn't be created. Less use of our precious limited natural resources.

This is a capitalistic country. We should continuously re-evaluate our priorities and determine where it is appropriate to have capitalism take a back seat to preserving our future. Common sense. If you are removing high quality resources at a fast rate and returning poor quality waste products to the earth, then eventually you will run out of high quality resources. It's very simple. It does not take a multi-year investigation to put the pieces together. But common sense is overlooked where capitalism and the quick dollar dominate and overshadow all reason. I've heard statements such as, "If I don't do this, then someone else will. Why should they profit and not me?" The well-thought-out process has been cast aside for the golden opportunity, regardless of the consequences. Only after individuals, business, and industry ventures have exploited the exploitable and the damage is done, only then do we start to ask questions. Our past mining and dam diverting practices are good examples. Manufacturing processes are another. Manufacturing and industry are good. So long as the process to produce the goods and services and the final products are WORTH the resources used and the subsequent effects on the environment. Business and industry have to take a hard look at what it produces and ask, "Is it worth it?" And consumers have to evaluate what they purchase, how the product made it to the shelf and ask the same question. Stepping outside yourself to view your actions from a global perspective takes more than most can handle. But that is true progress. Or we will be meandering through the maze for the next fifty years only with exponentially dwindling resources. Then the maze walls and floors collapse and it is only down from there.

Our populations, both U.S. and worldwide continue to increase. Increased populations put high demands on food and water supplies, and produce more wastes and require more septic systems and larger sewage treatment plants. Agriculturally, we are exhausting all research avenues that increase yield in agriculture. When the plant has reached it's ceiling on productivity and yield, then what?

If we irrigate more land to grow more food to feed growing populations, we use more irrigation water, fertilizer, and pesticides. The earth beneath our feet does not have the ability to replenish our water supply as fast as we are pumping it out of the ground for manufacturing, domestic, agricultural, and recreational use. Yet we continue to use our waters as though it were an unlimited supply. We have continued to use drinking water for uses where household gray water would have been much more water-wise. Many individuals don't see the problem. They turn on their faucets every day. Water comes out. They drive by lakes and rivers and see development in areas where none existed previously. There is no problem. But there is. Like a disease, you do not see the symptoms until the damage has begun.

Humans have been ingenious at coming up with various ways to produce more food and duplicate and synthesize food-related products. We have not yet devised a way of creating more water. All of our food-related progress is directly dependent on water. Water is the limiting factor. Ancient civilizations, sometimes too late, realized the importance of respecting the waters. (1)(12)(32) One would think that modern civilizations, with their advanced thinking and past knowledge base would have learned from the mistakes of ancient civilizations. Instead, driven by acquisition while positioning themselves above nature rather than part of it, advanced cultures have fallen short in their appreciation and preservation of water. The results are parched lands, communities running out of water, destroyed wetlands, contaminated waterways, and death. (1)(12)

The climatic conditions and location of bodies of water in this country makes the U.S. an Eden compared to other countries. Our waterways have been instrumental in tracing the movement of nomadic tribes, irrigating crops, carving our landscapes, trading goods with neighbors, contributing to our food supply, and in choosing locations for urban development, industrial, home, and recreational sites. Somewhere along the way, we lost our respect for the water resource. We worry about our oil dependence in this country, but at least there are alternatives. Whether or not we choose to explore those alternatives is another issue. Oil has been a powerful driving force in the human desire for power and the drive for land acquisition. But humans do not need oil to live. Oil makes our lives more convenient. Water makes our lives possible. If our water sources are insufficient or contaminated and can no longer meet our needs, what are our water alternatives? There are none. I can get by day-to-day without oil, but I can't get by without water. Our bodies are approximately seventy percent water. A one percent lowering of water in our body creates thirst, a five percent drop causes a slight body temperature increase, and at ten percent we can't function. A twelve percent loss of water and we die. (1)(10)(32) Knowing this, which reserves make you feel more secure, water or oil?

Globally, what other water sources can we tap? It has been suggested that we could remove the salts from ocean water and utilize that source. Desalination. Other parts of the world such as the Middle East do it. They have to. Saudi Arabia depends on the process for irrigation and dominates in the production of distillate water. Oil is normally used to power desalination plants. (32) The Middle East has oil, so it is a feasible solution. Each country has a unique set of resources available to them. Cultural traditions and influences, levels and types of industries, population issues, and governing bodies all heavily affect how the resources are utilized. Here in the U.S. we need to manage the waters we have rather than seek out and exploit more sources. When we truly focus on efficiency and re-use of resources and alternative energy sources, then maybe we can branch out and explore other water resources. Do we really want to drain the oceans? I hope not. What happens to that food resource? What about the aquatic life? The climate? Here, in the U.S., we need to use less, and re-use what we have. We, from state to state, country to country, need to learn from each other. Each region has limited resources in one or more areas. How each state and country handles it's deficiencies and solves each problem is an education for everyone. We need to appreciate how other countries handle their problems, observe, learn, and hopefully produce educated solutions to our own problems. Teamwork.

Water should not be bought and sold. It should not be paid for. It is a natural resource that every human needs access to. Food can be grown; water cannot. The allocation and ownership of water rights in relation to land acquisition from our ancestral history was a big mistake. No one should own a water source. The government allows private groups to profit by selling water to humans who require water to stay alive. To put a price on water restricts access to the poor. (1) It is every human's responsibility to protect, conserve, and efficiently use our water resources. It is also every human's right to access the same clean, potable water. There has been, are, and will continue to be water wars and closed door political meetings on acquisition of, control of, and distribution of water. (2) Until better solutions surface, we need to conserve what we have in every way possible.

One conservation concept is to re-use our water supply. Related to households, the re-use of grey (or gray) water is a crucial step in water management. Gray water is the water going into the sewer or the septic tank, except that which comes from the toilet (black water). For re-use purposes, gray water is the water from the shower, bathroom sink, bath tub or clothes washing machine drains. In most climates, and especially in drought-stricken regions, re-use of gray water can be the primary contributor of water for the growth and survival of a home landscape. Gray water normally contains phosphates from detergents, as well as dust, dirt, hair, dead skin cells, and body oils. Shower and laundry waters are acceptable gray waters. Diaper water, like toilet water, is not. Neither is kitchen sink water due to the meat fragments, oil, fat, and grease content. Not only are these substances hard to filter, but they are more likely to breed disease and attract pests. Due to disease issues, gray water cannot be collected and saved in catchments the way rainwater can. In order to use gray water, care must be taken to avoid using softeners, enzymatic cleaners, whiteners, and bleaches, as they contain certain chemicals that could result in the contamination of surface or ground water, and they are detrimental to plant and/or animal life. Health food stores carry an array of natural, biodegradable, non-toxic household cleaners that can be used in place of traditional household products. Gray water systems can be installed relatively quickly and easily at a reasonable cost. Many towns and counties still do not allow irrigation with gray water. With all plumbing systems running properly and the option of irrigating with gray water at one's fingertips, it is a crime to be irrigating lands with clean water. (17)

What happens to the gray water once it reaches the soil? The application of gray water has its limitations, and proper application is crucial to avoiding complications. Our soils have many functions, one of which is that of a filter. The soil with its' diverse community of beneficial organisms has a substantial capacity to filter and purify dirty water, so long as the soil is not inundated all at once. Allowed to stagnate, it will attract mosquitoes, flies, and rodents, and release offensive odors. The improper application of previously contaminated gray water can spread parasites, viruses, and bacteria, with the extreme examples being typhoid fever, dysentery, and infectious hepatitis. (17) Once systems are functioning and used properly, use of gray water is a wise solution for areas such as the southwest. In states and counties where it is legal to re-use gray water, there are several systems available designed to modify existing plumbing to carry out this task.

Replacing vegetation is another water-related conservation method. In the summer, lawn and garden care can use up to 80% of the home's total water use. Replacing high water use vegetation with low water use, drought tolerant and deep-rooted plants can help reduce water consumption. For example, tall fescue or creeping red fescue can replace traditional lawn and turf grasses since their root systems can extend down much deeper (to access soil water) than Kentucky blue grass. (7)(9)(17) Removal of unwanted, invasive, high water use trees is another example.

Using human excrement as fertilizer could be a conservation method, although it has not yet gained wide acceptance. This is an indirect viable water conservation method. There is a stigma attached to re-using human waste. Most individuals in the U.S. would prefer never to see this material again rather than re-use it. Additionally, knowledge that the improper processing of human waste can result in the spread of infectious diseases has kept the re-use of this resource from being fully explored. Re-use of excrement waste has always been a contentious and volatile topic. With our current method of using flush toilets, we water down wastes and flush the whole volume into a septic system and hope the microbial/bacterial life that breaks down these materials are doing their job properly. Or we pipe the materials to a treatment plant, separate the fluids from the solids and treat accordingly. Traditional septic systems are anaerobic (do not utilize oxygen) so little decomposition takes place there, and when it does, it is slow. There are several waste removal alternatives to a traditional septic tank. Toilet alternatives include vacuum flush models, low flush toilets, chemical toilets, incinerating (oil/bottled gas/electric) toilets, aerobic models, biological (utilize enzymes, bacteria, aerobic and anaerobic digestion) systems, and oil flush toilets. And there are alternatives besides traditional treatment plants and household septic tanks. Currently, there are pilot projects underway that focus on the immediate treatment and recycling of black waters. Pressure sewers, alternate separation and re-use systems, community bio-plants that include sanitary waste collection, community bio-plants in which wastes are transformed into methane gas and fertilizers, and self-contained living systems (components include anaerobic digesters, algal growth chambers, sedimentation chambers, sand beds, solar stills and gas exchangers) are all valid alternatives to traditional waste disposal. (17)(31) So why aren't they more popular? One reason is that there is no intentional direct effort to share information from one state to the next. Our water, conservation, and recycling issues are not all that unique. Many areas of several states have faced similar problems and issues. Some factors may be unique to a given area, but it is worth exploring what others have done in order to reap the benefits of their knowledge and avoid their mistakes. The budget, priorities, economics, and political agenda of a given area dictate the time, effort, money and emphasis placed on programs that address these issues. Therefore, some communities are many steps ahead of other communities. Another reason why many of these alternative systems are not commonplace is that they are not prominently marketed, seen in public areas, or widely promoted as part of construction and building options.

In order for alternative systems to become mainstream, we must:

1) Explore all our options and know everything about each operation, both positive and negative effects, prior to mass promoting and installation efforts. Some of our past mistakes have involved not thinking through all the possible cause and effect relationships. We need to involve individuals and groups who are able to see the entire picture clearly in order to avoid costly trust, health, and monetary mistakes. The bidding process should not be the determinant of who is in charge of a project. A track record of clear, concise, well-thought-out planning experience and an extensive knowledge base should be the determinants.

2) Initiate public acceptance. If local communities and businesses utilize these systems in their buildings and structures, individual acceptance will be much greater than if these systems remain as an obscure novelty item. Promote them in the visual media of television and movies.

3) Ensure public and health safety measures are taken. This goes hand-in-hand with exploring all options in 1) above as public and health safety must be of utmost importance. To

install a system county-wide only to find out shortly thereafter that the system imposes a health risk, not only creates a stigma attached to that particular system but also discourages the use of alternative systems in general in the public eye.

4) See government acceptance. Government acceptance and subsequent promotion and use of these systems in government buildings are necessary for private industry and individual acceptance. Incentives such as tax credits would encourage individuals to at least take a closer look at these options.

5) See working systems on display. Small-scale operations would be necessary in order to have a working model. To encourage understanding and education, working systems should be set up and left up across the country, at locations easily accessible to the general public.

6) Keep the cost of these systems low. Unfortunately, when a new product comes to fruition and is first available on the market, the consumer cost for that product is extremely high. Eventually, as competitors come into the picture the price is driven down. Consumers will not utilize alternative systems in their households or businesses if the initial cost or cost of maintenance is high. Efficient composting toilets are a good example. Although they have been on the market for many years, they have not yet been widely accepted. Part of the reason is cost. Several thousand dollars for one

toilet quickly encourages the homeowner/builder to look at more traditional and cheaper options such as regular flush toilets and septic tanks. Systems such as composting toilets are an investment. Costs must be kept low from the outset to encourage participation and use.

Household conservation methods include replacing regular toilets with low-flush toilets, replacing water flush toilets with composting toilets, replacing shower heads and faucets with energy efficient counterparts, installing re-circulating faucet units (to recycle the cold water that washes down the drain while waiting for the hot water), replacing dishwasher and clothes washers with more energy efficient models, not using the sink's garbage disposal, not washing the car often, using buckets of water and a sponge to wash a car rather than the hose, turning off the faucet while washing your face and brushing your teeth, installing irrigation timers and efficient irrigation systems for home, landscape, and garden use, and installing systems that enable the recycling of some household gray waters such as laundry and shower water. To those of you who actively take part in these methods and others, I commend you. Many of our citizens practice water conservation methods in their gardens, landscaping, and farming practices. These methods and others are mentioned through out this book.

In the southwest, individuals and communities are making a concerted effort to install low-flush toilets, re-use gray waters, and initiate water restrictions. These communities are taking a close look at or asking,

Where does the water come from?

Where does it go?

What is in the water when it goes back into the ground?

Who is using the water supply?

How much water are they using?

For what purpose are they using the water?

What else can they use in place of that water?

What systems can be installed to reduce water usage and still do the job?

Could they use less?

In asking these questions, we stimulate awareness. Unless we increase awareness and seriously change our attitudes, approach to, and use of our water supplies, substantially increasing numbers of the human population will feel the effects of the lack of fresh water within the next quarter century. (1)(2)(10)(17)(31)(32)

In reading this section, you may have come away thinking I am a pessimist, but I am not. I mentioned earlier that it was not long ago that we were pulling buckets of water up from hand dug wells. Look at how far we've come and how much we've learned from our ancestral past! We've created many problems for mother earth in the realm of exploitation of natural resources with regards to wastefulness, scarring of the landscape, soil, water, and air pollution and contamination, to name a few. I believe that collectively we can correct our mistakes to a great extent. Collectively. That includes individuals, small and big business, industry, community, and government. I say collectively because we all created these problems. Our need for goods means that more goods are then manufactured. And we all use goods. On that note alone, we are all responsible. Information and education are key to awareness of issues, the precursor to acting on the issues.

This is not a book about water politics nor is it a lecture on your water consumption habits. It is a book on awareness of the world around you, and particularly the world beneath your feet. Water in the soil in fluid form is a soil solution. When there is not enough of it in the soil to be fluid, it is in a vapor form or exists as a film on particle surfaces. Although not as obvious as lakes and rivers, it is extremely crucial to our existence.

3

SOIL COMPOSITION

Major Components

Soils are components of ecosystems along with microbes, animals, plants, geologic materials, climatic conditions, hydrologic conditions, and physiographic relationships. (5)(6)(16) Understanding the components of the soil aids in overall understanding of cause and effect as it relates to soil. That is what Chapter 3 is all about. If this chapter is hard reading, just breeze through it so you know what topics are covered then come back to it after you have read the remaining chapters. It will become clearer.

Soil is the unconsolidated cover of the earth that has been altered by pedogenic (natural soil formation) processes. It is made up of mineral and organic components, water and air, and is capable of supporting plant growth.

Soils contain solids, liquids, and gases. The solids are comprised primarily of sand, silt, and clay particles as well as organic matter. The arrangement of soil solids determines the amount of void or pore space that a soil possesses. (16)(25) Only clay and organic matter are of major importance in the nutrition of plants, since they are able to participate in chemical reactions. Gravel, sand, and silt are largely inert and contribute little to plant nutrition.

Soils develop due to three types of processes - physical, chemical, and biological (biophysical/biochemical ("bio" » living)). Physical processes include heating, cooling, freezing, thawing, wetting, drying, erosion, sedimentation, percolation, leaching, translocation, cryoturbation, and capillary rise. Chemical processes include hydrolysis, hydration, carbonation, transformation, oxidation, and reduction. Biological processes include all biophysical and biochemical processes involving plants, animals, and microbes. Soil formation is the result of combinations of these types of processes. (16)(25)

The seven factors responsible for soil formation include parent material (mineral and organic materials that are chemically, physically, and biologically weathered to form soils), climate, topography, plants, microbes and animals, time, and humans.

Soils are grouped as to the force that produced them: gravity, water, ice, or wind. Residual soils are those soils formed in place directly from the geologic material the soil rests upon. Colluvial soils are soils formed by gravity deposits (Think of rocks rolling down hill.) Alluvial soils are formed by stream deposits. Lacustrine soils are formed by lake deposits, marine soils, ocean deposits, glacial soils, ice deposits, and eolian soils, wind deposits. (18)

We tend to think of sands, silts and clays as separate materials where sand particles are gritty, silt particles have a powdery or talc-like consistency, and clay particles are sticky. That is correct, but there are also particle-size distinctions. There are 25 millimeters (mm) to an inch. Millimeters (mm) are the most common unit of measurement used for particle size. Sand (2-0.05 mm), silt (0.05-0.002 mm) and clay (<0.002 mm) designations can refer to either the size of the particles, where sands are the largest, and clays are the smallest or to the texture or feel of the particles. Additionally, clay-size particles are often confused with clay minerals. Clay minerals are certain minerals that, upon decomposition and weathering, produce clay (sticky and plastic) particles.

People often incorrectly associate the specific surface area of a particle to the size of a particle. Often it is thought that since sand particles are larger, that they have more specific surface area. This is incorrect. If we pretend that each particle is a perfect sphere, then a sand particle with a diameter of 2 mm (coarse sand size) has a surface area of 30 mm^2 /mm^3 . A silt size (0.02 mm) particle has a surface area of 3000 mm^2 /mm^3. However, for spherical particles of clay size, the surface area is 300,000 mm^2 /mm^3. The structure of a clay particle is that of a series of platelets piled one on top of the other. Since a clay particle is actually a stack of platelets, not a sphere, surface area per unit volume is about 10 times larger than 300,000 mm^2/mm^3. The surface area allows many chemical reactions to take place, which is why clays are so important to chemical processes in soils. (16)(30)

Clay particles are plastic and sticky when wet and are often hard and massive when dry. They adsorb (clings to the exterior and the faces of platelets versus absorb which refers to taking within) water, gas, and dissolved substances. Clay soils tend to collect water on the soil surface preventing water from moving freely through out the soil. Clays are commonly identified by their shrinkage abilities. If they swell when wet and shrink when dry they are called expandable clays, smectites, or montmorrillonites. If they do not shrink and swell they are kaolinitic or vermiculinic. Smectitic clay soils then can pose construction-related problems such as cracking foundations and roads. The platy structure facilitates trapping water in between the crevices. Some clay particles hold water so tightly that plant roots cannot access the water when needed. Surface cracking results in soil particles and organic material from the soil surface falling through the cracks and moving down the soil profile. Plants have difficulty getting established on these soils as there is too much movement for plants to stabilize their root systems. (5)(16)(18)

Clays are small, plate-shaped, alumino-silicate crystals, consisting of silicon, aluminum, iron, oxygen, hydrogen, and may also contain potassium, magnesium, and other elements. Two of the most important clays are kaolinite and montmorillonite. The montmorillonitic clays are smaller, have the ability to swell when wet and to shrink when dry, and are involved in physical and chemical reactions much more than kaolinitic clays. The terms cations and anions are usually mentioned in conjunction with soil chemical and physical reactions, so some definitions are in order. Ions are electrically charged atoms, groups of atoms, or compounds, the charge resulting from the loss of electrons (cations) or the gain of electrons (anions). Cations ("+" indicates positive charge) are positively charged ions whereas anions ("-" indicates negative charge) are negatively charged ions. Physical reactions include breakage and subsequent decomposition due to heating, cooling, freezing, thawing, wetting, drying, etc. Chemical reactions include transformations or changes that take place due to a variety of factors including but definitely not limited to ion exchange, additions or subtractions of materials such as organic matter, fertilizers, water, toxic materials, or exposure to air. Commonly, physical and chemical effects are intertwined. Let's use the pelting of raindrops on soft shale as an example. Shale eventually decomposes to form pliable, sticky clay. The action of the raindrop hitting the shale surface dents, chips and breaks away pieces of shale while the moisture itself facilitates transporting ions around and in shale crevices and pores and aids in chemical reactions that require moisture in order for the reaction to take place. In this example, both physical and chemical effects are taking place. (5)(16)

Cation exchange capacity (CEC) is the sum total of exchangeable cations adsorbed. It is expressed in milliequivalents (meq) per 100g of oven-dried soil. Most layered silicates are negatively charged. If the charged sites are not affected by pH they have a permanent or constant charge. If the ability to exchange ions (exchange capacity) increases with changes in pH it is a pH dependent charge. Organic matter has mostly pH dependent charge, as well as some constant charge. Montmorillonite has mostly constant charge and only some pH dependent charge. Vermiculite has all constant charge. Allophane has mostly pH dependent charge. Kaolinite, gibbsite, and goethite are primarily constant charge. Allophane, kaolinite, gibbsite, and goethite have positive charge when participating in anion exchange. (14)(15)(20)

Clay has the ability to adsorb cations. Clay platelets contain exchange sites where nutrients (as well as contaminants) are held and can be released into soil solution to be taken up by plant roots. Individual cations can be exchanged against each other. The hydrogen ions of the acids given off by plant roots can be adsorbed by the clay, while calcium, magnesium, and potassium are released in matching quantities into soil solution, aiding plant nutrition by being available in solution for plant uptake. (16)

Clay is a negatively charged colloid (very small organic and inorganic matter). This negative charge is the reason that positively charged cations surround each clay particle. Opposites attract. If the cations can get close to the clay surface, the negative charge is largely neutralized and the clay particles will cling together, an occurrence called flocculation. When calcium and magnesium are the dominant cations, flocculation commonly occurs. Soils with calcium clay generally have a more desirable structure than soils containing sodium clay. Sodium ions are large and are covered with a shell of water that prevents close interaction with clay particle surfaces enough to effectively neutralize clay's negative charge. Such clay particles therefore repel (called dispersion when due to sodium) each other and soils containing sodium clay are often dispersed. (16)(20)(30) Soils containing excessive amounts of sodium are referred to as sodic soils. Highways and roads built on sodic soils tend to collapse over time as the soil cannot hold together. This is one of the reasons that the practice of "salting" (which adds sodium) icy highways has been replaced with applying dirt, soil, or gravel instead.

The ability to exchange ions and retain certain nutrients is dependent on whether the clay minerals are of the hydrous oxide type, of the non-expanding lattice type, or the expanding lattice type. The layered silicate clay minerals have structures that are either 1:1, 2:1, or 2:1:1 layers of tetrahedral and octahedral sheets. Isomorphic substitution (the replacement of one atom by another of similar size in the framework without disrupting or changing the crystal structure of the mineral) occurs when one element substitutes for another in the clay mineral framework. If an element of a lower charge substitutes for an element of a higher charge, a permanent negative charge develops on the clay mineral. The arrangement, composition of, and sequence of the octahedral and tetrahedral sheets determine the expandability of the clay. One element can accept, release and replace another element in the mineral structure on exchange sites. High surface area and CEC give smectites the ability to adsorb cations. Clays, clayey soils, and organic matter commonly have a high CEC whereas sands and sandy soils have a low CEC. (5)(25)

Halloysite ($Al_2Si_2O_5(OH)_4 \cdot 2H_2O$) and Kaolinite ($Al_2Si_2O_5(OH)_4$) are 1:1 clay minerals and are low in isomorphic substitution. Halloysite is relatively unstable and tends to transform into Kaolinite over time. Kaolinite is commonly found in soils. (5)(6)(15)(20)

Illite (also known as hydrous mica) is a 2:1, nonexpanding, dioctahedral clay mineral. (5)(6)(14)(15)

Smectites (including Montmorrillonite (Al,Mg) $_8(Si_4O_{10})_4(OH)_8 \cdot 12H_2O$), Beidellite ($(Ca,Na)_{0.3}Al_2(OH)_2$-$(Al,Si)_4O_{10} \cdot 4H_2O$), and Nontronite ($Fe_2(Al,Si)_4O_{10}$ $(OH)_2Na_{0.3} \cdot nH_2O$)) are 2:1 clay minerals that have low to moderate isomorphic substitution and readily expand. (5)(6)(15)(20)

Vermiculite is a 2:1 clay mineral with high isomorphic substitution. Vermiculite has a high CEC. However, when vermiculite is saturated with K^+ (potassium) or NH_4 (ammonium) ions, it becomes non-expanding. (5)(6)(15)(20)

Chlorite ($(Mg,Fe)_3(Si,Al)_4O_{10}(OH)_2 \cdot (Mg,Fe)_3(OH)_6$) is a 2:1:1 clay mineral and is commonly found in many soils. Chlorites have a small surface area that reduces their CEC. (5)(6)(15)(20)

Sands and silts are rock fragments that have decomposed and weathered over time. They can consist of quartz, feldspar, mica, or a vast array of other minerals. Although sands and silts provide an anchor for plant root systems, hold materials in the soil, and aid in filtration, they are chemically inert. They are not responsible for the majority of chemical reactions that take place in the soil. That responsibility falls to the soil clay and organic matter components. Sandy soils allow fluids to infiltrate quickly, and often, too quickly. Very sandy soils tend to dry out frequently as there is not enough organic matter or clay particles to hold the moisture in between or onto the sand particles. Sandy soils are beneficial when drainage is necessary. Pore space affects soil moisture with regards to moisture's ability to reach plant roots, as well as affecting proportions of gas, vapors, water, and air.

Loamy soils are commonly considered ideal for plant growth as they contain relatively equal amounts of sand, silt, and clay. Loamy soils also tend to contain more organic matter, so these soils hold adequate amounts water, yet not too much water for surface pooling. The clay particles contain exchange sites where nutrients (as well as contaminants) are held and can be released into soil solution to be taken up through plant roots. As most growers of any vegetation know however, this generalization on loamy soils being "ideal" does not apply to all plants. Pure silt soils are uncommon as they are usually found mixed with other components. Silty soils are often found in areas where sediments accumulate.

For heavy clay soils, additions such as organic matter, cover crops, and sands incorporated into the soil will very slowly change the workability of these soils. Be warned however, that these soils often are very deep, extending down many meters. The severe cracking on the surface results in material on the surface falling down through the cracks, exposing more clay soil. One would have to apply many tons of organic material over many seasons to have a noticeable positive effect on these soils. Other soils may be shallower in depth and therefore easier to treat.

It is the feel of the soil when moist we are evaluating when we refer to soil texture. It is also the percentages of sand, silt, and clay that exist in a given sample. In the field, these percentages are estimated by feel, but are also sent to a laboratory where exact percentages are determined using particle size analysis methods (hydrometer and pipette). Using the three textural components of sand, silt, and clay, there are 12 textures of soil recognized by the United States Department of Agriculture (USDA) (28) including clay, sandy clay, sandy clay loam, clay loam, silty clay, silty clay loam, sand, loamy sand, sandy loam, loam, silt loam, and silt. The suffix tells you what dominates the texture. As you move to the left of the string, these components dominate less and less. So a silty clay loam would be very loamy in texture-loam, have a clayey feel-clay loam, yet would also feel smooth -silty clay loam. There are also modifiers added to these terms when certain percentages are met to indicate high percentages of a constituent. For example, if a loamy sand contains 25% or more very coarse and coarse sand and less than 50% of any other size sand, it is then called a loamy coarse sand. The percentages of sand, silt, and clay particles determine the textural class of your soil. For example, a soil that contains 40% sand, 40% silt and 20% clay is classified as a loam. (5)(6)(16)(28)

Soil also consists of rocks of various sizes, shapes, and composition. Gravels are 2 to 75 mm in diameter. Cobbles, 75 to 250 mm, stones, 250 to 600 mm, and boulders, greater than 600 mm in diameter. When they are flat and 2 to 150 mm long, they are called channers, 150 to 380 mm long, flagstones, 380 to 600 mm long, stones, and greater than 600 mm long, boulders. If there is 15-35% of some type of rock size in a given soil, the textural name is given a prefix of the dominant size fragment, for example, gravelly loam or cobbly loam. A soil containing 35-60% would be a very gravelly loam or very cobbly loam and more than 60% would be termed extremely gravelly loam or extremely cobbly loam. (28)

In addition to sand, silt, and clay particles and rock fragments, there is also organic matter, dust, remains of dead and decomposing biota such as animal carcasses, worms, insects, microbes, bacteria, and so on. Waste materials these creatures excrete and the chemicals they produce while living are included in this huge stock pot. Then there are human contributions to the soil, both as municipal waste as well as pollutants and contaminants. Knowing the composition of a soil helps to evaluate the ability of a given soil to retain soil moisture and how well it will respond to additions of amendments, and to treatments. Armed with this knowledge, actions can be taken to prevent moisture loss and conserve soil moisture as much as possible.

Salts

Major sources of soluble salts in soils include the weathering of primary minerals and native rocks, relic (or relict) fossil salts, atmospheric deposition, saline irrigation, road salt applications, drainage waters, saline ground waters, seawater intrusion, additions of inorganic and organic fertilizers, sludges, sewage effluents, brines from natural salt deposits, brines from oil and gas fields, and mining. As primary minerals in soils and exposed rocks weather, the processes of hydrolysis, hydration, oxidation, and carbonation occur, and soluble salts are released. Relic salts released from prior salt deposits or from entrapped solutions found in relic marine sediments are major contributors of salts in soils. (14)(30)

Soils naturally contain salts. Excessive salts are detrimental to both plants and soil. Pretend you have a glass of water with regular table salt stirred in. Taste it. It tastes salty. If you add more water to the glass, the solution will taste less salty. Remove a lot of water, the solution will taste very salty. The same analogy can be used for soil. As rocks, minerals and organic matter decompose, they release dissolved salts and nutrients into the water present in the soil. When there is enough liquid to be fluid, the water and the dissolved salts and nutrients are called a soil solution. When there is a lot of water relative to the amount or concentration of salt, the solution is dilute and plants may not be getting enough nutrients. When there is excessive salt relative to the amount of water, the solution is concentrated and plants will suffer or die. There are three major ways in which salts are detrimental to the plant-soil environment. Plants usually do not absorb salt and take it up through the plant root-that is not what commonly kills the plant. Salt increases the osmotic pressure around the plant root making it difficult to impossible for the plant root to absorb water (as well as nutrients since they are transported with the water). That is what injures or kills the plant. Increased osmotic pressure is the most detrimental effect of salts. The type of salt that dominates also affects the soil. If there is more calcium salt than sodium and magnesium salts, the soil will flocculate (good). If there is more sodium and magnesium salts than calcium salts, the soil will disperse (not good). Soils high in sodium have less pore space, create puddles when wet, and are difficult to work. Sodium salts break down the natural soil structure and generally create poor quality soils. Destroying soil structure is the second negative effect of salts. Salts of boron or lithium may become toxic and injure or destroy a plant before the effect of the increased osmotic pressure injures the plant. Toxicity is the third detrimental effect of salts. (6)(7)(16)(18)(20)(30)

Salt-affected soils are classified as saline, sodic, or saline-sodic soils. Briefly, saline soils are plagued by high levels of soluble salts, sodic soils have high levels of exchangeable sodium, and saline-sodic soils have both high levels of soluble salts and exchangeable sodium. Salt-affected soils occur most often in arid and semi-arid climates but they can also be found in areas where the climate and mobility of salts in solution cause periodic saline waters and soil. In arid and semi-arid climates, there is not enough water to leach soluble salts from the soil. An additional factor causing salt-affected soils is the high potential evaporation in these areas, which increases the concentration of salts in both soils and surface waters. It has been estimated that evaporation losses range from 50 to 90% in arid regions, resulting in 2 to 20 fold increases in soluble salts. The cations and anions of primary concern are Na^+(sodium), Ca^{2+}(calcium), Mg^{2+}(magnesium), K^+ (potassium), and Cl^- (chloride), SO_4^{2-}(sulfate), HCO_3^-(carbonate), CO_3^{2-} (bicarbonate) and NO_3^-(nitrate), respectively. (20)(30)

An important contributor to excessive salts in soils and waterways is the quality of the irrigation water used. If the irrigation water contains high levels of soluble salts, Na^+(sodium), B (boron), and trace elements, these excesses are inflicted on the soil-plant community. Saline irrigation water, low soil permeability, inadequate drainage, low rainfall, and poor irrigation management all cause salts to accumulate in soils which negatively affects plant growth. Affected soils must be leached with clean water to flush out salts. However the leaching of these salts can cause runoff and the pollution of bodies of water such as lakes, ponds, streams, and ground waters. (20)(30)

More than half of the earth's arable lands are located in arid and semi-arid regions that also tend to be salt-affected and/or alkaline. These soils have also played a unique role in world history as the rise and fall of many ancient civilizations were tied to the irrigation and subsequent mismanagement of alkaline and salt-affected soils. (5)(12) Rain and snowfall are insufficient in these areas (<500 mm annually) to leach out the base-forming cations (Ca_2^+, Mg_2^+, K^+, Na^+, etc) that are slowly released as the rocks and minerals weather. As a result, the percentage base saturation is high and pH values above 7 dominate. Poor internal soil drainage can result in minimal, if any, leaching of soluble salts such as NaCl (sodium chloride), $CaCl_2$ (calcium chloride), $MgCl_2$(magnesium chloride), and KCl (potassium chloride) creating saline and alkaline conditions. (5) Adequate drainage is needed to remove excess water from the plant roots for optimum plant growth, provide aeration, and to flush out excess salts. Soil moisture plays a key role in whether or not salts are present, flushed away, or exist in minor or major quantities.

Soil Geology, Chemical Composition and Properties

Primary minerals are those that formed during the cooling of molten rock and are predominantly silicate minerals. Examples of primary minerals in the order of those most to least resistant to weathering include quartz, muscovite, microcline, orthoclase, biotite, albite, hornblende, augite, anorthite, and olivine. Igneous rocks are composed entirely of primary minerals. Metamorphic and sedimentary rocks can contain abundant amounts of either primary and/

or secondary minerals. Secondary minerals are formed in the soil from soluble products derived from the weathering of rocks. Examples of secondary minerals in the order of those most to least resistant to weathering include geothite, hematite, gibbsite, clay minerals, dolomite, calcite, and gypsum. Clay minerals are one of the most important soil secondary minerals due to their expansive surface area and the ability to participate in chemical reactions. Carbonates and sulfates accumulate in sub-surface horizons and are the dominant secondary minerals present in soils in arid and semi-arid regions. (15)(25)

Surface soils generally contain in order of abundance in percent by weight of dry soil: oxygen, silicon, aluminum, iron, potassium, carbon, sodium, calcium, magnesium, titanium, hydrogen, manganese, nitrogen, phosphorus, sulphur (or sulfur), and other elements. (5)(6)(14)(16)

Mineral solubility, soil reactions, pH, cation and anion exchange, buffering effects, and nutrient availability are major chemical properties of soils. These are determined primarily by the nature and quantity of the clay minerals and organic matter present. (25)

As was mentioned earlier, anions are negatively charged ions and cations are positively charged ions. The ions themselves come from the weathering and decomposition of rock, organic matter, minerals, and bedrock as well as from sewage, waterways, fertilizers, toxic wastes, metals, etc. Ionic species may or may not change forms due to the chemistry in the solution surrounding them at the time. Many ions are exchangeable, meaning they are able to trade places with other ions (depending upon the size) on clay and organic matter particles. The ions that were replaced are then in the soil solution which is what is measured in such tests as pH (measures acidity and alkalinity) or electrical conductivity (measures salts). Soil reaction is the condition of acidity or basicity of the soil. It is often referred to as the soil's pH, which is commonly defined as the negative log of the concentration of hydrogen ions. Although we discuss soil pH, we are actually referring to the pH of the soil solution. If the soil contains a dominant number of hydrogen ions (H^+), the soil solution will be acidic, so we extrapolate back to the soil and say that the soil is acidic, which may not necessarily be true. Most crops thrive best in slightly acidic (6.5-6.8) conditions. Ground limestone ($CaCO_3$) is usually added to a soil to raise the pH. Most soils in the southwest have the opposite problem. Due substantially to the relatively high amounts of calcium carbonate, soils are commonly high in pH and often need to be lowered in order to grow healthy plants or a variety of plants. This has been accomplished in the past primarily with the addition of gypsum. Gypsum's formula is $CaSO_4^{2-}$ x $2H_2O$. Although gypsum has been used with some success on high pH soils due to the sulfate (or sulphate) (SO_4^{2-}) component of gypsum, it is not recommended except on soils that are high in sodium and considered to be "sodic soils". The reason for this is that adding gypsum also adds calcium to the soil, and in soils already high in forms of calcium, it is thought that additional calcium will aid in maintaining a high soil pH. This negative effect is overshadowed when gypsum is applied to sodic soils due to all the benefits of lessening the dispersive effects of the sodic soil. The benefits out-weigh the adverse effects.

Soil moisture is important with regards to pH due to most ions moving through soil as well as transported through plants via the moisture of a soil solution. At pH values up to 7.5, the dissolved carbonate species present is primarily carbonic acid (H_2CO_3). Between 7.5 and 10.5, the dissolved carbonate species present is primarily bicarbonate (HCO_3^-). Higher than 10.5, the dissolved carbonate species present is primarily carbonate (CO_3^{2-}). (14)(30) The main concern with regards to pH is the availability or unavailability of cations and anions, and the form of nutrients, toxic metals, and minerals available for plant uptake. Soil pH values below 5.5 indicate that soluble levels of several metals, particularly Al^{3+} and Mn^{2+}, may be high enough to be toxic to most plants and some organisms. When pH values are above 7 this is often indicative of very low solubility and availability of necessary micronutrient metal cations such as Zn^{2+}. Even more extreme pH values signify the presence of specific minerals or ionic forms. A pH above 8.5 is commonly associated with a high Na+ (sodium) content, whereas a pH below 3 is associated with the presence of metal sulfides. (5)(20)(30)

There are many cations and anions. Calcium (Ca^{2+}) is one of the more important cations. In acid soils calcium can improve soil structure and fertility. Sodium is a cation and is not well-liked in soils due to it's ability to destroy soil structure and cause soils to disperse or break down.

Hydrogen (H^+) is a cation and too much of it creates acidic conditions. Important anions include phosphate (PO_4^{3-}), sulfate (SO_4^{2-}), nitrate (NO_3^-), and bicarbonate (HCO_3^-), the former three of which are essential plant nutrients. Bicarbonate helps to break down soil minerals and bring plant nutrients into an available form. Phosphate is held tightly by soil colloids, adsorbs quickly and desorbs very slowly. Sulfate, nitrate, and bicarbonate are not adsorbed by clay and move freely with the soil water. (16)

Ions in bodies of water affect what is found in nearby soils. Common sources of sodium (Na^+) include halite (NaCl), sea spray, hot springs, brines, some silicates and rare minerals such as nahcolite ($NaHCO_3$). Common sodic silicates include plagioclase type albite ($NaAlSi_3O_8$), and nepheline ($NaAlSiO_4$). Most sodium results from natural ion exchange. Common sources of chloride (Cl^-) include halite (NaCl), sea sprays, brines, and hot springs. Common sources for potassium (K^+) include potash feldspar ($KAlSi_2O_6$) and sylvite (KCL). Common sources for calcium (Ca^{2+}) include calcite ($CaCO_3$), aragonite ($CaCO_3$), dolomite (($CaMgCO_3)_2$), gypsum ($CaSO_4 \cdot 2H_2O$), anhydrite ($CaSO_4$), fluorite (CaF_2), plagioclase, (anorthite, $CaAl_2Si_2O_8$), pyroxene (diopside, $CaMgSi_2O_6$), and amphibole $NaCa_2(Mg,Fe,Al)Si_{8-}O_{22}(OH)_2$. Common sources for sulfate, (SO_4^{2-}) include Pyrite (FeS_2), gypsum ($CaSO_4 \cdot 2H_2O$), and anhydrite ($CaSO_4$). Under some conditions, considerable quantities of sulfate may be obtained from organic sulfur compounds (e.g., combustion of coal and petroleum, smelting of sulfide ores, geothermal waters). A common source for magnesium, (Mg^{2+}) is dolomite

$((CaMgCO_3)_2)$. Magnesium also comes from the silicates olivine $((Mg,Fe)_2)SiO_4)$, pyroxene (diopside, $CaMgSi_2O_6$), amphibole $NaCa_2(Mg,Fe,Al)Si_8O_{22}(OH)_2$ and mica $(K(Mg,Fe)_3(AlSi_3)O_{10}(OH)_2)$. Common sources for carbonate, (HCO_3-) and bicarbonate, (CO_3^{2-}) include the atmosphere, sulfate reduction, calcite and dolomite. (14)(15)

Nutrient absorption by plants is usually referred to as ion uptake or ion absorption because it is the ionic form in which nutrients are absorbed. Ionic interactions may occur as cation-cation interactions, anion-anion interactions, or cation-anion interactions. (13)

Although a number of polyvalent cations display active roles in ion transport, calcium (Ca^{2+}) is probably the single most important one in sustaining the fabric of the absorption mechanism. (7)(22)(27) Calcium is an interesting cation in that it is generally non-toxic. Wide variations of calcium levels in the medium is common in nature without adverse effects provided excess levels of other ions that are toxic are not present. Calcium is the only cation that can occupy most of the exchange complex without toxic effects, especially in moderate to high exchange capacity soils. Many plants grow well in calcareous soils containing free calcium carbonate or gypsum or in soils containing 20-30 meq of exchangeable Ca^{2+}. Exchangeable Ca^{2+} can be used to release and remove some toxic exchangeable cations when the latter are present in excessive amounts. Calcium plays a crucial role both as a regulator of selective ion transport and also in maintaining the framework of membranes. (22) Levels of soil moisture can affect the availability of ions and the ion species present as well as their ability to be transported, participate in physical and chemical reactions, and be taken up by plant root systems.

Soil Structure

Soil particles that are held together by chemical and physical forces form stable aggregates (clusters of particles). Natural aggregates are often referred to as peds. Collectively grouped, the soil aggregates or peds form soil structure. Soil structure influences the amount of water that enters the soil (infiltration) and gas diffusion at the soil surface. Soil structure also plays an important role in the movement of liquids and gaseous substances through soils. Porosity is a function of soil structure. (25) A soil's structure refers to the arrangement of soil particles with respect to one another. If you were to look at a wedge of soil and gently pry off a section with the tip of a knife, what natural structure would be visible? Single grained, such as sand in a sand pile is "structureless". There are no cementing or binding agents to keep the particles together. Smaller particles fill the space between the larger particles. Soil particles cling together due to a combination of moisture, bacterial coatings, organic matter, and natural cementing or bonding agents. When these clusters have rounded edges to them, the structure is called "crumb", and resembles cookie crumbs. Crumb structure provides various size openings or pores in the soil and is beneficial for plant growth. Other soil structures include blocky (angular blocky and subangular blocky), platy, prismatic, and columnar. (28) These structures visually appear pretty much the same as the word used to describe them. Soil structure forms and is affected by many different factors including farming practices, the type of vegetation grown on the land, precipitation, irrigation practices, the geologic and mineral composition of the soil, and amendments. Soil moisture affects soil structure and a soil's structure can affect soil moisture.

Soil Pores

Large pores are readily drained by water and filled by air after a heavy rain. They create a valuable aeration system. Small pores hold water against gravity and pull up water from the water table by capillarity, and are therefore necessary in order to provide water to plants. The ideal structure is evident where large and small pores occur in a proportion that corresponds to the water and air needs of the plant in a given climate. In very humid or irrigated areas, large pores should dominate as water supply is adequate but the air supply might easily become deficient. Where natural precipitation is limited, and no irrigation is practiced, small, water-holding pores should dominate. Under ordinary farming conditions, in the humid temperate zone, the ratio between the volume of the large and small pores should be about 1:1. Total porosity is the percentage of the soil volume which is normally not occupied by solids. This space can be filled with either water or air. (16)

Soils hold water in pore spaces by the cohesive and adhesive nature of water and soil particle surfaces. Cohesive forces are the result of water molecule polarity and hydrogen bonding, which attract water molecules to one another. Adhesion forces are responsible for attracting water molecules to soil mineral and organic matter surfaces. These forces allow water to move upward in soils by capillary action, or along surfaces of soil particles as water films. Air takes up that part of the pore space not occupied by water. As the water content increases, the air content decreases. Plant roots require oxygen for their normal functions, just as the above-ground plant parts do and as animals breathe. In respiration, plant roots use oxygen and give off carbon dioxide. For this reason, soil air usually contains less oxygen and more carbon dioxide than that of the atmosphere. (16) A continuous replenishment is necessary to keep the oxygen content sufficiently high. Large pores and intermediate moisture content accomplishes this replenishment.

To Reduce the Loss of or Retain Soil Moisture . . .

1) Flush out excess salts from your soil.
2) Do not over-apply water to known salted areas.
3) Keep the soil solution pH neutral or tailored to the plants' needs.

4

ORGANIC MATTER

Organic matter (OM) can include grass clippings, farm manure, sod, all green manures, peat muck, crop residues, cottonseed meal, plant and animal residues, dried blood, dead animals, and commercial fertilizers, all in various stages of decomposition. OM influences soil physical, chemical, and biological properties and plays a crucial role in the chemistry of soils.

Enriching soil by adding OM improves aeration and soil fertility, reduces erosion, increases water-holding capacity, reduces runoff, increases CEC, improves drainage, buffers pH, increases the extent of root growth, increases soil aggregation, provides plant food, improves soil structure, furnishes energy for beneficial bacteria growth, improves porosity and infiltration, reduces the negative effects of pesticides and pollutants, produces carbonic acids, eases root penetration, and serves as a storage house for macro- and micro-nutrients. (18)

Soil OM content varies from less than 1% in sandy soils and soils of arid regions to almost all OM in poorly drained organic soils. Surface soils of farmlands commonly contain 2-10% OM. OM is a highly complex mixture of carbon compounds, nitrogen, sulphur, and phosphorus. Due to the continuous additions of plant, leaf, and twig litter, forest floors and swampy areas are usually comprised of fibric, hemic, and sapric organic materials. Fibric materials are the least decomposed of the three, contain high amounts of fiber, and are recognizable as to the plant material it originated from. Hemic materials are an intermediary between fibric and sapric and are barely recognizable as to the origin of the plant matter, yet are not as decomposed as sapric materials. Sapric materials are highly decomposed, no longer recognizable as to plant material, contain a low plant fiber content, and have a low water content when saturated. (5)

Humus is dark-colored soil organic matter, a colloidal substance highly resistant to decomposition. Humus contains about 50 percent carbon, 5 percent nitrogen and 0.05 percent phosphorus. Chemically, it is a combination of lignin, amino acids, and other nitrogen-containing substances. Lignin is the most decay-resistant constituent of the mature cell walls of plants, and amino acids are a component of proteins. Conditions favorable for humus formation are those present in moist and cool climates. Dry and hot climates do not promote humus formation.

Grassland prairie soils, due to decomposition of grass roots, often contain more humus than forest floor soils. Roots of the dense, fine, fibrous root system of grasses continuously die off and replenish, providing continuous additions of organic materials. Each year, in the U.S., depending on the region and climate, 25-50% of grass roots die off, providing sub-surface OM. The continuous additions of OM along with the extensive root system of grasses improve soil structure, water-holding capacity and overall soil quality. Soils in arid, semi-arid, and hot, humid regions commonly have less OM than soils in other regions. High temperature, wind velocities, and low precipitation in these areas discourage lush growth and encourage erosion. OM is made up of humic substances (humic acid, fulvic acid, and humin) separated based on their solubility, and biochemical compounds. Humic acids are soluble in bases, but not soluble in acids. Fulvic acids are soluble in both acids and bases. Humin is insoluble and is procured after humic and fulvic acids have been removed. Biochemical compounds include organic acids, proteins, polysaccharides, sugars, and lipids. (16)(25)

I cannot over-emphasize the benefits of adding OM to soil. OM has the ability to adsorb trace element pollutants such as lead (Pb), cadmium (Cd), and copper (Cu) and chemically bind these elements so they are not free to contaminate surface and sub-surface waters. OM adsorbs pesticides and other organic chemicals and can adsorb inorganic gases such as NO and NO_2, and organic gases such as CO. Due to adsorption, soils containing high amounts of OM require special attention when pesticides are applied. More pesticide needs to be added in order to overcome adsorption, yet too much will result in the excess being leached through the soil. Excessive applications then, can result in contamination of water resources. Most soils have the ability to adsorb and neutralize pollutants to harmless levels through chemical and biochemical means, so long as soils are not inundated all at once. There are limits, however. When combined with mineral particles, OM forms water-stable aggregates. OM conditions hard, clayey soil and provides cohesion to loose, sandy soil. In its decomposition, OM breaks down into minerals, water, and carbon dioxide. Carbon dioxide facilitates dissolving the soil minerals which converts them to a form available for plant root uptake. (16)

To think that reactions in the soil are due primarily to the sand, silt, clay, water, and OM components is a gross discredit to the life within the soil. Soil is alive with all sorts of organisms, both visible to the naked eye and microscopic. These organisms are crucial to the decomposition of organic matter, neutralizing the negative effects of contaminants, catalyzing chemical reactions, producing organic substances, adding organic materials, and promoting available plant nutrients. Soil organisms are often referred to as biota and include micro-organisms such as nematodes, protozoa, mites, spiders, insects, and earthworms as well as smaller organisms such as bacteria, fungi, and algae. Large animals such as cattle influence soils primarily through overgrazing, adding waste materials to soil, and compaction. Effects of overgrazing include reduced ground cover, altered plant species composition, soil compaction, and accelerated soil erosion. Animals of a smaller size that have a great impact on soil properties are those that burrow. Gophers, prairie dogs, badgers, moles, and mice are some of the burrowing animals that influence the development of soils by mixing topsoil and subsoil material. Tunnels and holes produced by burrowing animals increase aeration and water infiltration. Insects and subterranean animals move materials, and proceed to consume and excrete substances providing benefits to the soil. Many organisms have specific functions. Earthworms are a good example as they are visible and are extremely beneficial to soil quality. They ingest OM and soil which are decomposed by digestive enzymes and grinding action as they pass through the worm. Each day, the weight of soil material consumed and excreted (excrement and worm castings) by earthworms can equal their weight. Earthworm tunnels aerate the soil and break apart soil clods. Many of these activities occur in the topsoil. (25) Organic materials on the soil surface and mixed into the surface layer helps insulate the soil not only from wind erosion that would otherwise blow the top soil away, but by serving as a buffer to modify temperature extremes which helps protect soil organisms.

Precipitation and temperature affect both a plant's ability to thrive and the rate of OM decomposition. Moisture and temperature are crucial determinants of how active micro-organisms facilitate OM decomposition. Increased OM promotes a high water-holding capacity which, in turn, promotes vigorous plant growth and more OM. Soils that contain high amounts of OM hold moisture in the soil much more than soils without OM.

OM can be purchased if that is the only acquisition option available. Always ask for the lab analysis results (usually free) for whichever loose (versus pre-packaged) purchased compost you use. There are several companies that sell "green manure" such as grass and plant clippings in various stages of decomposition. To avoid bringing in weed seed or other undesirables, I would recommend composting the material before applying it to the soil. When composted correctly, the high temperatures usually kill off spores, weeds, and viable seeds that may be hiding within. For home gardeners, it is always a good idea to create a compost pile, not only to recycle vegetable scraps and plant material, but also to create your own supply of organic material. Although a compost pile started now will not be ready for use for at least a season or two, the benefits will be realized eventually. Depending upon the land area available to you, the compost area or pile can be as large (acres) or small (piles) as you want it to be. You'll find many combinations of the most efficient ways to compost at your local library. Composting can also be carried out in black plastic trash bags for small areas. The trash bags accelerate the process due to the higher heat in the bag. Be sure to poke holes in the bag or leave it loosely tied or gases may build up to such an extent that your bag explodes! Composting bins and fancy equipment are also available to purchase as are bags of OM, but this is the most expensive option. Regardless of which option you choose, start adding OM now to improve and enrich your soil and water assets. A healthy soil utilizes water much more efficiently than an unhealthy soil.

To Reduce the Loss of or Retain Soil Moisture . . .

4) Add organic matter.
5) Create a compost pile.

5

SOIL MOISTURE

Plants require water. Water is a plant food. Water serves as both a solvent and a transporter for substances necessary for plant growth, substances of which are absorbed by the plant root when dissolved in solution. Water is the transport vehicle for the distribution of nutrients from one part of the plant to another. Structural plant cell walls need water in order to retain rigidity and succulence. Transpiration from plant roots to plant leaves cools the plant similar to the cooling effect of human perspiration. (18)

Our dependence on soil moisture is tremendous. Soil moisture is required by plants. Humans consume plants directly, or indirectly by the consumption of meat. Soil moisture filters our soils of pollutants and toxins, and aids in maintaining the stability we need to build structures, roads, etc, to maintain our environment. Soil organisms require moisture to carry out their functions, both physical and chemical. Soil moisture aids in erosion control and reduces the loss of topsoil by wind. We require soil moisture in order to maintain human existence on earth.

A water molecule contains one oxygen atom and two smaller hydrogen atoms. The elements are bonded together covalently, each hydrogen atom sharing its single electron with the oxygen. The hydrogen atoms are attracted to the oxygen in sort of a v-shaped arrangement at an angle of 104.5°. The shared electrons are closer to the oxygen than the hydrogen, creating an electronegative charge. The side on which the hydrogen atoms are located tends to be electropositive. Molecules such as water, whose positive and negative charge centers are off-balance are polar molecules. Due to polarity, each water molecule does not act independently but is linked with other water molecules. The hydrogen (positive) end of one molecule attracts the oxygen (negative) end of another, resulting in a chain-like network. Polarity is responsible for some important properties of water. Due to polarity, water molecules are attracted to electrostatically charged ions and to colloidal surfaces. Cations such as H^+, Na^+, K^+, and Ca^{2+} are attracted to the oxygen (negative) end of water molecules. Likewise, negatively charged clay surfaces attract water at the hydrogen (positive) end of the molecule. Polarity of water molecules also encourages the dissolution of salts in water since the ionic components have a greater attraction for water molecules than for each other. (5)

When the air above ground is dry, moisture on the soil surface evaporates into the air. As the soil surface air evaporates, moisture is drawn up through the soil through capillary action, where it then evaporates at the soil surface. Capillary action refers to a solution's ability to move up a tube or passageway. Many factors affect these naturally formed passageways by which water and nutrients move up through the soil.

Soil water is available in three forms, hygroscopic, capillary, and gravitational. Hygroscopic water is, basically, not available to plants. Soil particles (specifically clays) have hygroscopic water trapped within and around the particle layers that plants are unable to access. Capillary water is movement from one soil particle to the next, within the film of water itself. Two forces cause capillarity: 1) the attraction of water for the solid walls of channels through which it moves (adhesion or adsorption)) and 2) the surface tension of water, which is due largely to the attraction of water molecules for each other (cohesion). Capillarity can be demonstrated by placing one end of a thin glass tube in water. The water rises in the tube. The smaller the tube opening, the higher the water rises. Water molecules are attracted to the sides of a tube and creep up the tube in response to this attraction. The cohesive force between individual water molecules pulls water up the tube. Unlike glass tubes, soil pores are irregularly shaped. Some soil pores are filled with trapped air, slowing down or preventing the movement of water by capillarity. Capillarity is commonly thought of as an upward movement, but it can occur in any direction. Soil films play a crucial role in water flow in drier climates such as the southwest as films are often the only water available. Gravitational water is in excess of capillary water and tends to move downward in direction. (5)(15)(18)(23)

Infiltration is the process of water soaking into the soil. The infiltration rate is measured in inches or millimeters per hour (in/hr, mm/hr). Good infiltration is when water easily enters the soil where applied. Medium-textured soils with a high OM content have good infiltration rates. Sandy soils have high to excessive infiltration rates. Where the soil is compacted, slopes, clods, contains heavy clays, is frozen, or has poor tilth due to poor management practices, water will not easily infiltrate, and will instead pool or result in runoff. Runoff contributes greatly to erosion, reduces water quality in water bodies, and is an inefficient use of water. Additions of OM will increase the water-holding capacity and slow the infiltration rate when incorporated in sandy soils, and will increase the infiltration rate in heavy, clayey, and cloddy soils.

Water moving down a soil profile continues moving out of the soil and eventually reaches a zone of water saturation that lies above a layer of rocks or clays. The upper portion of this zone of saturation is the water table, or commonly, our ground water. Ground water can frequently be found 1-10 meters below the soil surface in humid regions but may be hundreds or thousands of meters deeper in arid regions. The unsaturated zone above the water table is frequently called the vadose zone. Downward-percolating drainage water reaches the shallow ground water, continues through porous geologic materials (aquifers), replenishes deeper ground water and eventually reaches springs, streams and rivers, or it is pumped for human use. The water table level fluctuates in response to the amount of water percolating through the soil. In humid temperate regions, the water table is commonly highest in the spring following winter rains and snow melt, and before hot summers take over. Ground water is a significant source of water for potable domestic, industrial, and agricultural use. Shallow wells commonly provide water for irrigating crops or for rural dwellings. Deep wells supply water for municipal, industrial, and agricultural purposes. Due to the close proximity of the surface of shallow wells to land applications of fertilizers, pesticides, wastes, etc, deep wells are preferred for a cleaner drinking water supply. More layers to filter through implies a cleaner, safer drinking water product. While drainage water from soils is a major source of recharge for these underground water resources, it is insufficient to compensate for the rate of human use of these waters in some deep reservoirs. The rate of water removal far exceeds the rate of recharge. Consequently, these deep water tables are becoming deeper. Near ocean coastlines, as the water is removed by pumping, seepage of saltwater into these caverns (called saltwater intrusion) is a frequent occurrence. (5)(14)

The energy level of soil water is affected by three forces. Adhesion is one force. It is the attraction of soil solids (matrix) for water and is responsible for matric force (applies to adsorption and capillarity). The second force results from the attraction of ions and other solutes to water, resulting in osmotic forces. Gravity is the third major force and it is responsible for pulling water downward. The theory behind gravitational water is that the energy level of soil water at any one elevation in the soil profile is higher than that of pure water at some lower elevation. It is this difference in energy levels that causes water to flow in a downward direction. (5)

Water moves through soils either as a vapor or a liquid. Vapor flow through a soil is generally a slow process. Water vapor is present in unsaturated soils and moves by diffusion within the soil due to vapor pressure and temperature gradients. There is internal and external water vapor movement in soils. Internal movement takes place within the soil pores. External movement occurs at the soil surface where water evaporates. Water vapor will move from a moist soil where the soil air is nearly 100% saturated with water vapor (high vapor pressure) to a drier soil where the vapor pressure is lower. Salt lowers the vapor pressure of water so water will move from a zone of low salt content to one with a higher salt content. Temperature affects both the movement of water vapor as well as vapor pressure. If the temperature from one part of a uniformly moist soil is lowered, the vapor pressure will decrease and water vapor will move toward this cooler part. Heating will increase the vapor pressure and the water vapor will move away from the heated area. In the southwest, vapor water movement is of considerable significance especially in supplying water to drought tolerant desert plants, many of which can exist at an extremely low soil water content. (5)(7)(9)(23)(24) Soil water movement is either saturated or unsaturated flow depending on the soil moisture content. Saturated flow occurs in soils where the void space is filled with water, and unsaturated flow occurs whenever void spaces are partially filled with air. Water flow within soil is a function of the driving force acting on the water (called the hydraulic gradient) and the ability of the soil to allow water movement (called hydraulic conductivity). The capillary fringe is the area above the water table where water in small pores is drawn upward by capillary action. (25) Water flow takes place in the direction of decreasing potential. Most of the processes involving soil-water interactions and root-zone water of most crop plants occur while soil is in an unsaturated condition. (5)(11)

Following rainfall or irrigation, some of the water will drain downward quickly in response to the hydraulic gradient (mostly gravity). After one to three days, this rapid downward movement almost ceases. The soil is then at field capacity. At this time, water has moved out of the macropores (large pores), and is replaced by air. The micropores (small pores) or capillary pores are still filled with water and continue to supply plants. Water movement will continue to take place, but it is unsaturated flow, and therefore, very slow.

In the process of plant roots absorbing water, they reduce the soil moisture content, which reduces the water potential in the soil in the immediate area surrounding the roots. In response to this lower potential, water moves toward the plant roots. Root depth then becomes significant in order to capture as much moisture as possible covering a larger soil area. During periods of hot, dry weather, when water demands and evaporation are high, capillary movement is an important means of providing water to plants. (5)(7)(27)

Cycling water from the earth's surface to the atmosphere and back again is environmentally crucial and is driven by solar energy. About 1/3 of the solar energy that reaches the earth is absorbed by water on or near the earth's surface. The absorbed energy promotes evaporation of the water, and the resultant water vapors move up into the atmosphere, forming clouds. Eventually, pressure and temperature differences in the atmosphere cause the water vapor to condense into liquid droplets or solid particles which return to the earth as rain or snow.

Evaporation is the process of converting liquid into a vapor state. Transpiration is a specialized type of evaporation in which the conversion from the liquid to vapor state takes place in the plant leaf. Usually, the two processes are grouped together (when referring to plants) and referred to as evapotranspiration. (25) In evapotranspiration, moisture on the plant leaf surface evaporates to the dry air. More moisture is then drawn up from the soil solution around the plant root, up through the plant to the surface of the leaves, where it evaporates again. The plant draws moisture up from the roots, not only to keep the plant structure moist, succulent and rigid, but also to provide nutrients (contained in the soil solution) to the plant. The soil water, then serves as a transport mechanism to get the nutrients where they belong, and deposit the water of the cargo to the plant leaf surface. This is a continuous process, the rate of which increases in hot, dry weather. In humid or wet weather conditions, the rate of evapotranspiration slows considerably due to the atmospheric deposition of dewy moisture onto the leaves as well as the rest of the plant. The plant gets a "rest" from work. The driving force is the environmental condition surrounding the exterior of the plant. If conditions are dry, more water will be demanded of the plant. If conditions are wet, less water will be demanded. Ever notice how vibrant, alive, and succulent plants look after a rainfall or after being misted in a greenhouse? Although most of the plant moisture is wicked away at the plant leaf surface, the moisture content of all plant parts above the ground surface is also depleted as the plant "bakes" under extremely dry conditions. Under very dry conditions sustained for long periods of time, the plant is unable to transfer the water and nutrients up through the plant fast enough to keep up with the rate of evapotranspiration. You'll notice that the leaves are commonly the first plant part to show signs of wilting. A sustained period of time under these stressful conditions that weakens the plant to such an extent that the plant does not recover to thrive and reproduce is called the permanent wilting point. (5)(6)(16)

From a physical point of view, evapotranspiration can be viewed as a continuous stream flowing from a periodically replenished source of soil moisture, to a sink of virtually unlimited capacity (though of varying evaporative potential)-the atmosphere. As long as the rate of uptake of soil moisture balances the rate of loss by transpiration, the stream continues while the plant remains hydrated. When the uptake rate decreases below transpiration, the plant begins to suffer. This imbalance results in loss of turgidity and in wilting of the plant. (7)(11)

The available water-holding capacity is a term used to describe the water that is stored in the soil and available for uptake by plants. Not all water in the soil is available to plants. Some water is held so tightly in some clay particles, for example, that plants cannot access it. The amount and rate of water uptake by plants depends on the ability of the roots to absorb water from the soil with which they are in contact, as well as on the ability of the soil to supply and transmit water toward the roots at a rate sufficient to meet transpiration requirements. This is dependent on the properties of the plant (rooting density, rooting depth, the expanse of root extension, and the physiological ability of the plant to continue extracting water from the soil at a rate needed to avoid wilting), and the properties of the soil (hydraulic conductivity-diffusivity-matric suction-wetness relationships). These relationships are also dependent upon the meteorological conditions (dictate the atmospheric need, which then draws from the plant and soil). (11)

Evaporation of soil water from the soil surface results in not only the loss of water but also the danger of soil salting, as discussed in Chapter 3. Evaporation from the soil surface can be controlled by modifying the albedo (soil color value) through color or structure changes of the soil surface, by shading the surface, by lowering the water table, or warming the soil surface so as to promote a downward-acting thermal gradient, or by changing tillage and mulching methods. Another approach to evaporation control is to induce a temporarily higher evaporation rate so as to rapidly dry out the soil surface to arrest or retard subsequent outflow. Shallow cultivation practices that crush the soil at the surface may cause the loosened layer to dry faster and more completely while helping to conserve the moisture of the soil below. Whether or not this method works consistently is still up for debate. (5)(11)(18)

Soils are often classified as wet, moist, or dry. A wet soil contains water that is held with little or no energy and will drain off by gravity quickly if the soil has enough large pores to permit drainage. While the soil is wet, it creates an anaerobic (air deficient) condition which is detrimental to most plants as well as to most beneficial microbes. The wet range extends from saturation to field capacity. (5)(16)

A moist soil contains sufficient water and air in proportions that are conducive to the growth of crop plants and microbes. Moist soil is friable (easily crumbled) and can be worked by machinery without destroying structure. Soil water in the moist range moves from wetter to drier locations in the soil, no longer due to gravity, but to capillarity. Soil water in the moist range is also called capillary water. The moist range extends from the field capacity to the hygroscopic point. At this point, soil is too dry to support plant growth. Before the dry limit of the moist range is reached, the water is held so tightly that plants cannot uptake water fast enough to balance the loss by transpiration and consequently, the plants wilt. This moisture condition (near the dry limit of the moist range) is therefore called the wilting point. The amount of soil water between field capacity and wilting point is the available water. (16) Plants that reach permanent wilting point usually will not fully recover, if at all, when adequate moisture is applied.

When the soil is dry, it still contains some water, but this is held with such a force that plants can no longer make use of it. Fine-textured soils (clay loams, silty clays) are hard in this condition and resist the penetration by tillage implements, especially if the soils are deficient in organic matter. Dry soils have the ability to absorb moisture from the atmosphere if the vapor pressure gradient is favorable (if the atmosphere has relatively more moisture than the soil). For this reason, soil water in the dry range is called hygroscopic. To measure the amount of water in a soil, a sample is weighed, dried to a constant weight in an oven at 105° and weighed again. The difference in weight is attributed to moisture. So a soil that weighs 10 grams moist and 7 grams when dried has a water content of 30% by weight. (5)(16)

Water escapes the soil surface mostly due to evaporation, seepage, transpiration, evapotranspiration, runoff, and inefficiency in application. Areas prone to collect excessive water (usually due to location and slope) should be either diverted to other soil areas, re-distributed to cover other vegetated surfaces, or channeled to other waterways in need of additional water.

To Reduce the Loss of or to Retain Soil Moisture . . .

6) Cover water bodies.
7) Improve water-holding capacity by adding OM.

6

IRRIGATION

Home gardening water consumption is called watering whereas watering crops is referred to as irrigation. In this book I use both interchangeably to mean the same thing-the watering of plants. In the United States, approximately 40 percent of the total water use is for agriculture, so it makes sense to use the most efficient watering system and methods available.

The quality of the water added to a garden or landscaped area can affect soil properties as well as plant growth vigor. If the water is high in salts, then you are adding salt to your soil. If you suspect your water to have a high salt content, have your water tested. Call your county extension agent, your local water department, or your local university (if they have a soil and water testing laboratory) to find out who tests water, what they test for, cost, and who can interpret the results of the test for you. The same salt issues apply to fertilizers-many have a surprisingly high percentage of salts. Read the back of the package carefully, or call your county extension agent to see what he/she knows about that particular fertilizer. There are filtration systems and devices to install in irrigation systems if salty irrigation waters are a problem.

The evaporation rate in irrigated areas of dry, hot, and arid regions is very high. Controlling evaporation and evapotranspiration is difficult, but some irrigation practices are better suited to minimize moisture loss than others. Surface, sprinkler, and micro-irrigation systems are the most common methods used to apply irrigation water. Each has their advantages and disadvantages. With surface systems, water is commonly diverted to the field in canals or ditches where rocks, concrete, or plastic are used to line the canals. This is the least efficient in use of the irrigation water due to evaporation loss enroute to the fields in addition to the wastefulness of widespread applications of water to an area versus more selective application methods. Sprinkler systems of center pivot design send water outward from a central point. Self-propelled traveling sprinklers are common as well and are usually found irrigating large areas of farmland. Water use efficiency in sprinkler systems is higher than surface systems. There is some loss due to evaporation while in flight, and some of the spray is carried away from its mark. The advantage is in the cooling effect of water on the plant itself (reducing transpiration rates) as well as the cool temperature of the droplets by the time they have touched the ground. Sprinkler irrigation systems should be tailored to apply water at rates that never exceed the water's ability to infiltrate the soil. (11) The use of trickle or drip irrigation has been found to be effective. Through this system, water is dripped or trickled out of pipes or tubes alongside each plant with the water targeting the plants, not the soil between the plants. This system has proven to be very efficient due to specifically targeting plants and has less evaporation loss compared to other systems. In some operations, perforated tubes through which water can flow are buried several centimeters below the soil surface and immediately under the row of plants. These micro-systems are designed to minimize evaporation loss from the soil surface. (5)(23)(32) Ideally, irrigation lines would be placed under the soil surface but above the roots so the water would be available to the plant root system. Holes in the lines that would release water would be placed directly above the root system that requires watering. One major drawback to earlier versions of sub-surface irrigation is that the roots were drawn to moisture and would quickly wrap themselves around any such water-producing line in the ground, clogging the holes, restricting flow, and damaging the plant itself (by cutting off its own water supply) in the process. The second major drawback was the maintenance of keeping the lines free and unclogged while not damaging the plants. Modern sub-surface systems have minimized or eliminated many of these problems.

Placing irrigation lines above ground results in moisture loss due to evaporation, wind, and irrigation spray patterns (by carrying moisture away from the intended land area). The compromise then is to place irrigation lines above the soil but below a gravel or rock layer. Additionally, an indentation around each plant forming a bowl shape will help catch and direct rainfall to the plant root zone. This is often not feasible with field crops. Drought conditions create a desirable environment for some pests that prey on weak plants. To help protect trees and plants from these pests, be sure the soil area within a 3-4 inch radius of the base of the plant is not disturbed by the bowl. In other words, start the bowl about 3-4 inches away from the plant base. In this way you are encouraging the spreading out of the roots. Watering cones are also available for individual plants as a way of delivering water below ground to the roots rather than applying it to a soil surface where it will evaporate.

Timing and Frequency of Watering

Many parts of the southwest implement water restrictions in order to conserve the water resource to ensure an ample supply in the future. The time of day you choose to water could be the difference between moist soil and healthy plant growth versus dry, baked surroundings. The time of the day of watering plays a key role in the soil's ability to retain moisture as well as the plant's ability to prepare for the hot dry days. When you water during the peak sun periods, much of the moisture is evaporated into the dry air immediately. More frequent watering is then necessary in order to keep plants from wilting and the soil from completely drying out. Watering during the day on sunny days results in a great loss of moisture to evaporation and evapotranspiration. The rate of evapotranspiration is highly dependent on how dry or moist the air is surrounding the plant itself as well as the plant's ability to draw moisture up from the roots. Under windy conditions the rates of evaporation and evapotranspiration increase substantially.

With watering restrictions in place, follow the restriction guidelines as to what day of the week and usage quantities your town, city, county or state imposes. However, you must use common sense. Use only as much water as you need, and no more. Your needs will depend on many factors including those that are represented in all of the chapter headings in this book. If you utilize all the information contained within these chapters, your water need will be low and efficiently utilized. You have to pay your water bill. Everyone suffers if you use more than what you need. If you can meet the demands of your household or business with 5,000 gallons, why use 10,000? Over-use is wasteful, careless, greedy, and short-sited.

Water either during the evening when the sun is going down or as early in the morning as possible (preferably before the sun brightens the sky). Watering during these times gives your plants a chance to "refuel and stock up" to better equip them to handle the stress of the daytime hours. Watering during these time periods also gives clay and OM particles a chance to hydrate which can increase the amount of water available to plants during the hot, dry day. Less water is lost due to evaporation and evapotranspiration during these periods. If you water during the day, the plant is wilted by the time the sun goes down and has no resources to draw on in order to recover. Clays and organic matter are two of the most common soil constituents that absorb and release moisture. Some of this stored moisture is available for plant roots to utilize, and some is not. Moist OM particles aid in facilitating microbial activity and in keeping the ground moist and the ground temperature cooler.

To Reduce the Loss of or to Retain Soil Moisture . . .

8) Use efficient irrigation methods.
9) Irrigate in early morning or evening hours.
10) Have water tested if salts are suspected.
11) Excavate a bowl around plants whenever possible.
12) Collect and save rainwater in sealed/covered catchment containers.

7

SOIL TEMPERATURE

Soil temperature fluctuates both during day and night hours in a 24 hour period as well as with each season. Temperature fluctuations are minimized with increase in soil depth. Sub-surface soils are generally warmer than the atmosphere in the winter and cooler in the summer. The release of stored winter energy and moisture during the spring and summer months is what helps maintain a constant temperature in the subsoil. At about three feet in soil depth, there is no difference between day and night temperature. At twenty feet, the temperature remains the same regardless of the season. Where the soil is covered by a dense growth of vegetation or a thick layer of mulch, temperature variations are much less severe and do not penetrate as deeply. Soil temperature has a direct effect not only on plant growth but also microbial activity, water movement, and soil structure. In addition to liquid water movement, water vapor movement is affected by temperature differences in the soil or between the soil and the air interface. Since it takes more energy to heat water than to heat air, a cold soil is a soil that retains much of it's water in the spring, and warms up slowly as compared to the same soil with air filling the pores. (16)(18)(22)(25)

The temperature of a soil directly and indirectly affects chemical reactions and physical and biological processes occurring in that soil. In cold soils, rates of chemical and biological reactions are slow. Decomposition by microbes is substantially decreased, slowing the rate at which plant nutrients such as nitrogen, phosphorus, sulfur, and calcium are made available. Absorption and transport of water and nutrients by plant roots are also slowed by low temperature. Each plant species has different needs for germination and optimum growth. Reproductive cycles, seeding and fruiting are accelerated, delayed, or halted as changes in soil temperature triggers chemical messages that mobilize plants to shift from one phase in plant growth to the next. (5)(16)(18)(22)(25)

Only 35-40% of the solar radiation actually penetrates our atmosphere to reach earth in cloudy humid regions, whereas 75% reaches the earth in cloud free arid regions. Incoming solar energy is spent primarily to evaporate water from soil or leaf surfaces, or is radiated or reflected off surfaces back to the atmosphere. Only about 10% is absorbed by the soil. (5)

The angle of the sun to the horizon affects soil temperature. If the sun is directly overhead, the incoming path of the rays is perpendicular to the soil surface, so energy absorption and soil temperature are increased. Predictably, how much shading a given land area receives will affect soil temperature and soil moisture as shaded areas are cooler in temperature than open areas. Land that slopes south absorbs more heat than any other direction. Soils located on sun-facing slopes warm more quickly in the spring, have the advantage of earlier plant growth, and the disadvantage of drying out more quickly. (5)(18)(24)

Mulch, soil moisture, and soil surface color have significant effects on soil temperature. In periods of hot weather, a mulch layer keeps the soil cooler than where no mulch is used. During cold spells in the winter, mulches moderate rapid temperature declines, so they serve to buffer extremes in soil temperatures. Alternately, frost penetration and damage is greater in bare, non-insulated soil. A dry soil is more easily heated than a wet one due to the amount of energy required to raise the temperature of water by 1 degree Celsius (it's heat capacity). It is much higher than that required to warm soil solids by 1 degree Celsius. Or, put another way, it takes about five times as much heat to raise the temperature of a pound of water one degree of temperature as it does to heat a pound of dry soil one degree of temperature. When water becomes warm it evaporates and in doing so removes much heat from soil. Soils containing a lot of water will be cooler than soils containing less water. Dark soil surfaces tend to absorb radiation in the form of heat, whereas the lighter colors tend to reflect radiant energy. This should not necessarily imply however, that dark soils are always warmer. Soils that are dark due to having higher amounts of organic matter hold larger amounts of water. Water requires more energy to warm up and serves to cool the soil as it evaporates. Consequently, sandy soils are warmer than clayey soils. (5)(18)

We don't normally think about what factors affect soil temperature or what effects soil temperature has on soil moisture. If the soil surface is covered with light-colored pebbles, stones, cobbles, or boulders, these materials will reflect more light and absorb less heat than dark-colored pebbles, stones, cobbles, or boulders. Likewise, light-colored soils are known to reflect light and deflect heat where dark-colored soils absorb. Light-colored materials then, will result in cooler soil temperatures. Cooler soil temperatures will result in less soil moisture loss due to evaporation. If you are in a drought-prone part of the country such as the southwest, light-colored materials covering the soil surface are preferred in order to minimize soil moisture loss. However, a layer of dark colored stones in the spring facilitates ground warming, so dark materials have their advantages. The labor involved in stone placement may not be labor and time feasible for many people. (18)

Adequate drainage is necessary to facilitate soil warming. In the spring a wet soil may be up to 8 degrees cooler than a drier soil. Draining excess water lowers the specific heat of the soil, resulting in quicker warming of the drained layers. Concurrently, the cooling effect of surface evaporation may be reduced, warming the soil. So, water can be used as a temperature control method.

Soil and plant materials weather or break down faster under warmer environmental conditions. Soils in tropical regions, due primarily to the combination of steady heat and moisture, tend to break down and weather at a rapid rate. Microbes help break down organic and inorganic particles as well as catalyze chemical reactions. Microbial populations increase substantially under ideal heat and moisture conditions. A home compost pile is a good example to witness the effects of heat and moisture. Compost piles work most effectively when the right amounts of adequate moisture, carbon-containing materials, and nitrogen-containing materials are present. Layering the materials in a particular order, adding sufficient water, allowing the mass to sufficiently heat up, and occasionally turning the pile to aerate, leads to a proper breakdown of the compost material. This material can then be applied to soil to improve the texture, tilth, infiltration, and water-holding capacity as well as contribute nutrients. Decomposition in a compost pile is relatively fast. It is easy then to visualize tropical soils and the fast rate at which materials break down.

Air surrounding plant roots is usually lower than air temperatures during the growing season, and rapid fluctuations in the temperatures of the root zone are less than that of the air to which the tops of the plants are subjected. As a result, the roots of plants have become less adaptive to temperature extremes, and are more sensitive to sudden changes. The roots of most plants would die if exposed to the same conditions of temperature to which the tops are subjected. The rate of vertical root growth into the lower (colder) horizons (soil layers) is slower than horizontal root growth. In the early stages of plant growth the roots in the surface horizons will be more branched and numerous than the roots extending into the lower horizons. Temperature differences and delicate, easily broken root systems are partly why transplanting plants should be done in such a way to minimize disturbance to the roots. (7)(24)

Factors that affect soil temperature include the air temperature, the intensity, quality and duration of radiant energy, precipitation, soil moisture, evaporative potential of the air, color and thermal conductivity of the soil, and surface cover. These factors also directly or indirectly affect plant growth.

To Reduce the Loss of or to Retain Soil Moisture . . .

13) Use water as a temperature control method.
14) Apply mulch.
15) Plant on slopes that maximize temperature/moisture resources.
16) Use rock fragments to absorb or reflect radiation in the form of heat.
17) Adjust soil surface colors.
18) Mulch thickly.

8

SOIL COLOR

Soil color is often an indicator of past or present events of the soil, and the presence or absence of minerals and forms of minerals. Soil color aids in distinguishing horizons, identifying soil properties, processes, and features and also facilitates the classification of a soil. Soil color under field conditions is affected by many factors including the quality and intensity of the light, surface reflectance characteristics of the soil, moisture content of the soil sample, pigmented clay coatings on clay particles, relief of the soil surface, the degree of crushing of the soil sample, and the individual's eyesight and perceptions. (3)(21)(29) Soil colors can include red, orange, yellow, green, blue, violet, black, brown, and white as well as shades of many hues in between. Soils are not usually comprised of particles of one color but of particles of many colors mixed together that form a dominant hue. By definition, hue is the attribute of a visual sensation according to which an area appears to be similar to one, or two properties of two, of the perceived colors red, yellow, green, and blue. Value is the attribute of a visual sensation according to which an area appears to exhibit more or less light. Chroma is the attribute of a visual sensation according to which an area appears to exhibit more or less of it's hue. (4)(19)(26) The main effect on plant growth is that various colors absorb and reflect light and heat waves differently and therefore results in differences in temperatures with the corresponding changes in soil moisture, structure, microbial activity and decomposition. Whenever the soil is covered by plants or mulch, effects due to the soil surface color cease to exist.

Light gray and blue colors indicate the leaching of iron and other cations. Red colors suggest an accumulation of iron, commonly resulting from the leaching out of silica. Mottling (groupings of blotches) of certain colors indicates seasonal excess of water in the soil resulting in zones of oxidation and reduction. Most soil colors are derived from the colors of iron oxides, organic matter, and the silt and clay coatings that coat the surfaces of soil particles. Organic coatings tend to darken and mask the colors derived from the iron oxides. Subsoil horizons with little organic matter, therefore, often most clearly display the colors of the various iron oxides that coat the soil particles. Some examples of these oxides are the yellow of goethite, the red of hematite, and the brown of maghematite. Other minerals that commonly give soils distinctive colors are manganese oxide (black), and glauconite (green). Carbonates typically accumulate in the soils of semi-arid regions in the southwest and are usually whitish in color. Prolonged, very wet, anaerobic conditions, (such as those conditions found in swamps or marshes) can cause iron oxide coatings to become chemically reduced, changing the red or brown color to gray, greenish, or bluish colors, a condition referred to as gley. (3)(5)

To Reduce the Loss of or to Retain Soil Moisture . . .

16) Use rock fragments to absorb or reflect radiation in the form of heat.
17) Adjust soil surface colors.

9

MULCH

Any material applied to the soil surface to keep weeds down and minimize evaporation may be called mulch. Plastic sheets, landscaping fabrics, crop residues, rocks, gravel, soil, straw, bark, leaves, sawdust, and manure all qualify. Mulches are highly effective in reducing evaporation and are most practical for the home garden and for areas containing specialty plants or crops. Plant growth in arid regions such as the southwest is not sufficiently high to provide adequate residues to sustain the organic mulch level needed to reduce evaporation, hence the need to apply mulch from other sources. Specifically prepared paper and plastics are options as mulches. This cover is laid out either between the rows or over the rows. Black plastic or other sheet materials can be used to effectively control weeds, provide a micro-watershed effect, and serve as an evaporation barrier. Black films intercept solar radiation and convert it to a sensible heat, most of which is then re-radiated. Clear plastic allows most of the solar energy to pass through to the soil, where it is converted to sensible heat and trapped. As long as the ground is covered, evaporation from the soil surface and encroaching weeds are minimized. Unless the rainfall is very heavy, the paper or plastic does not seriously interfere with the infiltration of rainwater into the soil. The cost of the materials and the difficulty of keeping it in place are primary considerations in utilizing these options. Furthermore, the temperature of the soil under the plastic is commonly 8 to 10 degrees higher than with a straw mulch. While this may be helpful for plants established early in the spring, it may be too much heat for temperature-sensitive plants. Soil temperatures below mulches in temperate climates are usually lower (with the exception of plastics) and water contents are usually higher than where mulches are not present. Applying surface mulch often results in better water conservation and moisture distribution in the germination zone of soils. (5)(24)

The sun's energy that is received at the ground surface heats soil, evaporates water, heats air, and is reflected off surfaces back into the atmosphere. The more dense the vegetative or mulch cover, the less moisture will be lost due to evaporation from the soil surface.

One of the most widely accepted conservation tillage practices that produces a mulch is stubble mulch tillage. In this method, crop residues from the previous crop are re-distributed on the soil surface. The land is then tilled in such a way to ensure that residues remain on or near the surface. Wheat stubble, straw, and cornstalks are commonly used. With the no-tillage method, the new crop is planted directly amongst residues of the previous crop with no plowing or disking. Till-planting is where residues are swept away from the immediate area of origin, but most of the soil and the surface area remain untilled. Till-planting leaves most of the soil covered with crop residues to help reduce evaporation and water vapor losses from soils. (5)

One of the early debates that still continues today about the control of evaporation is whether or not the formation of a soil mulch is a desirable moisture conserving practice. Creating a loose, mulch layer of soil on top of the soil is thought to break capillary movement of water upwards and therefore conserve soil moisture loss due to evaporation. The soil mulch interrupts capillarity by breaking the cohesion of particles to each other. Mulching has the effect of disrupting the capillary movement of soil solution up through the soil. If you disrupt the capillary movement, let's say in the top 6 inches of your soil the surface 6 inches now serves as a covering or dressing for the soil below. Additionally, you sacrifice the baking of the top 6 inches in order to protect the soil below, conserving moisture in the process. And lastly, you disrupt and cut off the capillary movement of water and soil solution upwards where it would evaporate at the soil surface. Likewise, shallow cultivation, which crushes the soil at the surface, has the effect of causing the loosened layer to dry faster and more completely and may help conserve the moisture of the soil below. Some studies show this mulch method is ineffective in humid regions and is most effective in regions with distinct wet and dry seasons, as in the tropics. (5)

Pebbles, gravel, cobbles, stones, and boulders (all referred to a fragments) can also be used as a mulch and are effective in soil moisture conservation. Pebbles are 2 to 75 mm in diameter. Cobbles are 75 to 250 mm, stones, 250 to 600 mm, and boulders are greater than 600 mm in diameter. When they are flat and 2 to 150 mm long, they are called channers, 150 to 380 mm, flagstones, 380 to 600 mm, stones, and greater than 600 mm, boulders. (5)(6)(28) Used as a mulch and placed close together, rock fragments are effective in reducing evaporation as well as absorbing, radiating, and reflecting heat and light based on the color, density, and geologic properties of the fragments.

As the earth cooled, the crust which was once magma, cooled and solidified. This solidified material is igneous (examples include granite and basalt) by origin. As the igneous rock was broken down in solution, formed a sediment, then re-solidified into rock (commonly under pressure), the sedimentary rocks were formed. Examples of a few of these include limestone, shale, sandstone, and dolomite. Limestone is relatively soluble in water. Shale is commonly a hardened clay whereas sandstone is compressed or cemented sand particles. Metamorphic rocks were formed by heat, pressure and other processes acting on either igneous or sedimentary rocks. The properties of metamorphic rocks are commonly completely different from that of the material from which they were derived. Gneiss is a metamorphic rock derived from granite. Unlike granite, it has a layered structure due to pressure. Marble is metamorphic and derived from limestone or dolomite with heat and pressure. It is crystallized and not as soluble as limestone. Slate is metamorphic and formed from shale with heat and pressure. It is harder than shale. (15)(18) Gravel and rocks work well to allow rainfall to penetrate between fragments to the soil. The more fragment surface in contact with the most soil surface area results in substantial decreases in losses due to evaporation. Think of a gravel the size and shape of a quarter. The gravel laying face down on the soil will reduce evaporation losses much more than the gravel standing on edge. Additionally, one large flat rock the size of a football will reduce water loss to evaporation much more than many smaller flat rocks laying with sides touching. The cracks separating the smaller rocks are allowing moisture to evaporate, whereas the single large rock has no such cracks. So the size and shape of the gravel or rock layer matters. Alternately, the smaller gravel allows more water to be more evenly distributed. The ease of handling, placing, and maintaining these materials is a primary factor to consider when using these materials as mulch. (8)

Some organic polymers applied to the soil surface aid in the rapid formation of a crust. Commonly, there exist physical crusts, biological crusts, and chemical crusts. A physical crust is composed of soil material that has decreased pore space and increased density. These properties create a dense layer that separates itself from the soil surface. A biological crust is a living, respiring, collective assembly of lichen, moss, cyanobacteria, and algae thriving on the soil surface, providing chemical compounds and physical structures that bind the mass together. We tend to protect areas containing biological crusts as these are living organisms that add nitrogen and other nutrients and substances to the soil below as well as provide habitat for other life. Additionally, biological crusts may provide a resource for future medicinal needs. A chemical crust forms when there is a high salt content at the soil surface. Some chemical applications can increase the angle of contact between the water and the soil, and some can result in the formation of a crust. Crusts may create resistance to evaporation from the soil surface as well as cut off the capillary action of water movement upwards through the soil. The disadvantage to crusted soils is that water infiltrates with difficulty and is unevenly distributed over a given area.

To Reduce the Loss of or to Retain Soil Moisture . . .

18) Mulch thickly.
19) Mulch using an inexpensive material that you are willing to maintain.
20) Use rock fragments to decrease evaporation from the soil surface.

10

SOIL EROSION

The top layer of the soil is what protects the underlying soil from moisture loss, so what happens when the top layer is removed? Soil erosion is the natural process of removing materials from one area and placing it elsewhere. Erosion can occur from the movement of water carrying particles away, from heavy rains pelting the soil surface hard enough to dislodge soil, or by wind where particles are carried away. Careless applications of irrigation waters and misdirected drainage waters result in runoff, which in turn, can create eroded gullies. Human activities such as the clearing of land of it's natural vegetation will result in an exposed surface prone to erosion. Clayey soils subject to surface cracking often facilitates organic material on the soil surface falling into the cracks leaving the surface prone to erosion. Sediments washed away from one area are usually deposited somewhere else downstream. Terms, such as "silting up" or "sediment build-up" are commonly used in reports for highway or dam construction projects. These terms refer to the fine particles that are easily transported from one place to another either by water or wind. Materials put down as mulch or a layer to protect the soil surface will help prevent soil erosion and concurrently protects the moisture contained within the soil. The climate, steepness (percent slope), direction of water movement (aspect), and surface cover will dictate the amount of loss due to water and wind, and where these materials might end up being deposited. Erosion removes the binding substances that hold particles together (especially erosion caused by rainfall impact), fills beneficial soil air space, removes nutrients and organic materials, destroys the soil's natural structure, creates a barren landscape, exposes root systems, removes valuable topsoil, and makes it difficult for plants to establish themselves. Additionally, erosion wastes precious water by taking it away from needed areas and inundating other areas in a short period of time. Sediments clog drainageways, streams, waterways, ditches, roads, and dams. Forceful winds claim much of our world's topsoil each year. Topsoil is the surface layer of soil with the greatest amount of organic matter (OM), biological activity, and nutrients. Tree, wall, or vegetation barriers can be set up to minimize soil losses due to wind. Barriers are also used to protect plantings while the plant roots get established.

Applying irrigation water only to where it is needed in a volume that does not promote erosion and runoff is a crucial step in reducing soil moisture loss and retaining the water resource.

To Reduce the Loss of or to Retain Soil Moisture . . .

21) Keep lands covered with vegetation or erosion control barriers.
22) Be sure waters are flowing where they are supposed to be flowing.
23) Use mulch as an erosion control measure.

11

PLANTS

Green plants are comprised of approximately 80% water, 14% carbohydrates and fats, 4% protein, and 2% minerals. There are five primary functions crucial to plant life:

Absorption of water and nutrients by the roots.

Transpiration of water from the plants (mostly the leaves).

Photosynthesis-the creation of plant material through the chemical combination of the carbon dioxide of the atmosphere, the water of the soil, and uv light. The following synthesis requires green chlorophyll as the catalyst.

$$6 \ CO_2 + 6 \ H_2O + uv \rightarrow C_6H_{12}O_6 + 6 \ O_2$$
carbon dioxide + water + uv light → sugar + oxygen

Synthesis of complex organic compounds. Carbohydrates, fats, proteins, lignin and many other compounds are formed from simple sugars, nitrogen compounds, and minerals (salts)(requires respiration).

Respiration. Plant tops and roots inhale oxygen and exhale carbon dioxide. Respiration is often thought of as the reverse process of photosynthesis:

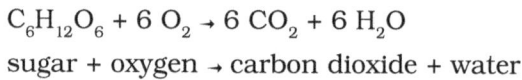

$$C_6H_{12}O_6 + 6\ O_2 \rightarrow 6\ CO_2 + 6\ H_2O$$
$$\text{sugar + oxygen} \rightarrow \text{carbon dioxide + water}$$

Except for oxygen utilized in respiration for the above-ground parts and the carbon dioxide, most of the other plant nutrients comes from the soil. These include water, oxygen (utilized in the respiration of the roots), and minerals. (5)(6)(16)(20)

Plants draw quantities of water from the soil in excess of their essential needs. In dry climates, field plants may consume hundreds of tons of water for each ton of vegetative growth. Plants utilize about 200 to 1000 pounds of water in exchange for one pound of dry matter. This water requirement varies with the site and climate. In fertile soil in a moist climate, where growing conditions are good, the amount of water needed to produce one pound of dry matter is much less than in a poor soil with inadequate moisture supply. Well over 90% of the water plants extract from the soil is transmitted to an insatiably thirsty atmosphere. Plants can thrive in an atmosphere saturated or nearly saturated with vapor and therefore require very little transpiration. Transpiration is caused by the vapor pressure gradient between the normally water-saturated leaves and a dry atmosphere, especially in the southwest. The evaporative demand of the climate in which plants live dictate the transpiration rate. Plants can limit the rate of transpiration by shutting the stomates of their leaves. However, the same stomates which transpire water also serve as receptacles for the uptake of the carbon dioxide needed in photosynthesis in order for the plant to grow. Limiting transpiration could result in limiting plant growth. Additionally, reduced transpiration can lead to warming of the plants and an increase in respiration rate.

The rate of water uptake for a given volume of soil is substantially affected by rooting density (the effective length of roots per unit volume of soil) and soil conductivity. If the soil moisture is uniform throughout all depths of the rooting zone, but the active roots are not uniformly distributed, the rate of water uptake will increase where the density of the root mass is greatest. (11)(16)

Phreatophytes were mentioned in an earlier chapter but are worth mentioning again. Undesirable phreatophytes are plants and trees that extract water from the soil layers above ground waters, from waterbodies, stream channels, irrigation and drainage canals, and floodplains, reproduce at a rapid rate, take over land areas driving away other native plant species, and pump excessive amounts of water from the ground and into the atmosphere.

Phreatophytes pose a problem all over the U.S. and are a serious problem to areas such as the southwest because water and drought issues are always of major concern in these areas. Some of the least desirable and most common phreatophytes include salt cedar (Tamarix pentandra pall.), willows (Salix spp.), cottonwoods (Populus spp.), mesquite (Prosopis juliflora), and greasewood (Sarcobatus vermiculatus). Cutting down these vegetative species will only result in sprouting and re-growth. Therefore, eradication is necessary. Undesirable shrubs and trees should be removed from the root and burned. Recyclers can chip the wood into mulch or it can be burned as firewood.

Transpiration by weeds can extract soil moisture equal to and in excess of that used by valued plants. Since weeds take water away from the soil as well as from beneficial plants, weeds contribute to soil moisture loss, so weed control should definitely be a conservation method. A weed is not a specific group of nasty plants but rather any plant you do not want growing on your landscape.

Tree and shrub barriers that protect a land area from wind and provide shade can be beneficial to wind prone areas by reducing water loss due to evaporation and wind erosion of surface soil. Additionally, the barriers shade and provide cooler temperatures to reduce moisture loss attributed to warmer temperatures. If a soil is irrigated and produces a high yield (yield can include vegetative cover, fruits, vegetables, or flowers), that is an efficient use of irrigation water. If a soil is being irrigated and has a low yield, that is an inefficient use of irrigation water. That being said, the use of fertilizers to increase yield is an efficiency measure in water conservation. Alternately, stop growing plants that with adequate moisture, consistently produce low yields, as that is an inefficient use of water. All the fertilizer in the world won't benefit your plants in the absence of sufficient moisture. All efforts should be made to choose plants indigenous to the area. (9) Resist the effort to bring in exotic plant species that may use more water and could bring in more problems in relation to pests and disease. Xeriscape and choose deep rooted plants. Planting only plants that have dual functions such as food and erosion control would be ideal. People are often swayed into planting exotic or high maintenance plants because they look different than what their neighbor has. Are the added cost, care, maintenance, disease, pest, and water issues worth it? Xeriscaping with native plants, as well as combining mulch practices and pavements can be beautiful and more water efficient than most other landscapes and is ideally suited to the southwest. Your local nursery can advise you on which plants are best suited to your area.

To Reduce the Loss of or to Retain Soil Moisture . . .

24) Remove unwanted phreatophytes.
25) Pull out all weeds.
26) Strategically place wind/shade barriers.
27) Xeriscape.
28) Plant native and deep-rooted plants.

12

MISCELLANEOUS

Topics in this section are too short to devote an entire chapter to, but this should not detract from the value of the information contained herein.

Try not to step on, dig in, or run heavy equipment over your land when it is wet. The soil will become compacted, destroying the soil's natural structure and with repeated treatment, will become hard and difficult to work. Compaction lowers the soil's water-holding capacity and decreases the ability of water to infiltrate. Normally, this advice is reserved for farmers who tend to use heavy tractor and tillage equipment, but it also applies to the homeowner.

Collect and save rainfall in underground storage tanks or in covered, sealed catchment storage systems. Using catchments to collect rainwater is an ancient practice that has not lost it's value in modern times. Collected water can then be distributed only where needed. Receptacles must be securely covered not only to keep out insects and debris, but to discourage mosquito breeding grounds and the diseases (such as the West Nile Virus) they spread.

Cover ponds whenever possible to reduce evaporation loss to minimal. Cover bare soil areas with vegetation or mulch materials to reduce evaporation from the soil surface.

Only apply water resources to those soil areas that are most productive and that have the highest potential to be productive. It is inefficient to continue to expend time and water to care for plants that do not thrive. If the problems cannot be corrected, it is best to do something else with these areas. Fertilizers are often used to increase production. In this respect, fertilizers are seen as a means to efficient water use. As I've stressed in this book, a healthier soil results in more efficient water use. Soils can often benefit from a "jump-start" of inoculants such as fungi, algae, and bacteria, as well as additions of humus and proteins (to increase micro-organism activity). Crude proteins are commonly derived from waste products of ocean life and provide a food source to facilitate microbial health and reproduction. Inoculants are referred to as root bio-stimulants. Root bio-stimulants can increase the overall health of the soil as well as increase yield. Frequently, when inoculants are added to soil, fertilizer use can be decreased. Additions such as those just described should be considered as part of the long term outdoor management effort. Costs of adding amendments to soils require careful consideration to find the balance between low costs, high yields, low maintenance, and maximum water-use efficiency.

Strategically design your outdoors to place structures and vegetation to provide shade, block winds, and aid in erosion control.

Create natural vegetative mats whenever feasible by using plant materials as ground debris. Twigs, bamboo, leaves, and moss are just a few examples of vegetative mat materials. Stepping on these mats not only helps to break down and work the organic material into the ground, but also helps to retain soil moisture while providing a natural path to walk upon. Waste supermarket and restaurant vegetable scraps as well as household vegetable matter can be recycled to a compost bin, or, when feasible, chopped in a blender to produce partially broken down materials. These materials can then be used to create a mat. Blending vegetable scraps in the blender and pouring it outdoors creates an interesting woven mat from the vegetable fibers. As the liquid is drawn down into the soil, the vegetable matter dries, forming an interwoven (so long as the material was not pureed) crust (interesting addition to a compost pile). Beware, however, as it is messy and also attracts insects.

In the southwest around moist areas, termites are sometimes a problem. So when you are designing your outdoors, be sure your plantings and water management methods are set up in such a way to keep moisture, irrigated vegetation and any moisture-loving vegetation as far away from structures as possible.

To Reduce the Loss of or to Retain Soil Moisture . . .

29) Do not compact soil when it is wet.

30) Collect rainwater for re-use.

31) Cover ponds and bare soil areas.

32) Design your outdoors.

33) Irrigate only those plants with the highest potential.

34) When adding inoculants, fertilizers, or other amendments to soil, find balance between low cost, high yields, low maintenance, and maximum water-use efficiency.

35) Keep moisture and plantings away from structures to avoid termite problems.

13

TIPS TO REDUCE THE LOSS OF OR RETAIN SOIL MOISTURE

1) Flush out excess salts from your soil. Ch. 3, 6

2) Do not over-apply water to known salted areas. Ch. 3, 6

3) Keep the soil solution pH neutral or tailored to the plants' needs. Ch.3

4) Add organic matter. Ch. 4, 7

5) Create a compost pile. Ch. 4, 12

6) Cover water bodies. Ch. 5, 6, 12

7) Improve water-holding capacity by adding OM. Ch. 5

8) Use efficient irrigation methods. Ch. 6

9) Irrigate in early morning or evening hours. Ch. 6

10) Have water tested if salts are suspected. Ch. 6

11) Excavate a bowl around plants whenever possible. Ch. 6

12) Collect and save rainwater in sealed/covered catchment containers. Ch. 6, 12

13) Use water as a temperature control method. Ch. 7

14) Apply mulch. Ch. 7, 10, 12

15) Plant on slopes that maximize temperature/moisture resources. Ch. 7

16) Use rock fragments to absorb or reflect radiation in the form of heat. Ch. 7, 8

17) Adjust soil surface colors. Ch. 7, 8

18) Mulch thickly. Ch. 7, 9, 10

19) Mulch using an inexpensive material that you are willing to maintain. Ch. 9

20) Use rock fragments to decrease evaporation from the soil surface. Ch. 9

21) Keep lands covered with vegetation or erosion control barriers. Ch. 10, 12

22) Be sure waters are flowing where they are supposed to be flowing. Ch. 10

23) Use mulch as an erosion control measure. Ch. 10

24) Remove unwanted phreatophytes. Ch. 11

25) Pull out all weeds. Ch. 11

26) Strategically place wind/shade barriers. Ch. 11

27) Xeriscape. Ch. 11

28) Plant native and deep-rooted plants. Ch. 11

29) Do not compact soil when it is wet. Ch. 12

30) Collect rainwater for re-use. Ch. 12

31) Cover ponds and bare soil areas. Ch. 12

32) Design your outdoors. Ch. 12

33) Irrigate only those plants with the highest potential. Ch. 12

34) When adding inoculants, fertilizers, or other amendments to soil, find balance between low cost, high

yields, low maintenance, and maximum water-use efficiency. Ch. 12

35) Keep moisture and plantings away from structures to avoid termite problems. Ch. 12

REFERENCES

(1) Barlow, M. and T. Clarke. 1111. Blue gold: the fight to stop the corporate theft of the world's water. The New Press, New York.

(2) Baxter, J.O. 1927. Dividing new mexico's waters 1700–1912. Univ. of New Mexico Press, Albuquerque.

(3) Bigham, J.M. and E.J. Ciolkosz, (eds.) 1993. Soil color. SSSA Publication #31. SSSA, Madison, WI.

(4) Bourgin, D. 1999. http://www.well.com/user/rld/vidpage/color_faqhtml.Color spaces FAQ P.1–24.6/24/99

(5) Brady, N.C. and R.R. Weil. 1996. The nature and properties of soils, 11th ed. Prentice Hall, Inc., Upper Saddle River, New Jersey.

(6) Buol, S.W., F.D. Hole, and R.J. McCracken. 1989. Soil genesis and classification, 3rd ed. Iowa State Univ. Press, Ames.

(7) Carson, E.W. (ed.) 1974. The plant root and its environment. Proceedings of an institute sponsored by the Southern Regional Education Board, held at Virginia Polytechnic Institute and State University, July 5–16, 1971, The University Press of Virginia, Charlottesville.

(8) Cox, J. 1999. From vines to wines; the complete guide to growing grapes and making your own wine, Storey Books, North Adams, Mass.

(9) Dunmire, W.W. and G.D. Tierney. 1995. Wild plants of the pueblo province; exploring ancient and enduring uses, Museum of New Mexico Press.

(10) Gottlieb, R. 1988. A life of its own: the politics and power of water. Harcourt Brace Jovanovich, Publishers, San Diego, New York, London.

(11) Hillel, D. 1982. Introduction to soil physics, Academic Press, Inc., Harcourt Brace Jovanovich, Pub., San Diego.

(12) Hillel, D.J. 1991. Out of the earth; civilization and the life of the soil, The Free Press, A Division of Macmillan, Inc., New York.

(13) Hiatt, A.J. and J.E. Leggett. 1971. Ionic interactions and antagonisms in plants, in The Plant Root and Its Environment, Proceedings of an institute sponsored by the Southern Regional Education Board, held at Virginia Poltechnic Institue and State University, July 5-16, 1971., edited by E.W. Carson., University Press of Virginia, Charlottesville.

(14) Hounslow, A.W. 1995. Water quality data; analysis and interpretation. Lewis Publishers, Boca Raton

(15) Klein, C. and C.S. Hurlbut, Jr. 1977. Manual of mineralogy, 21st ed., John Wiley and Sons, Inc., New York.

(16) Kohnke, H. 1986. Soil science simplified, Waveland Press, Inc., Prospect Heights, Illinois.

(17) Kourik, R. 1988. Gray water use in the landscape: how to help your landscape prosper with recycled water. Metamorphic Press.

(18) Lyon, T.L. 1926. Soils and fertilizers. The Macmillan Co., New York.

(19) MacAdam, D.L. 1985. Color measurement, theme and variations. Springer-Verlag, Berlin.

(20) McBride, M.B. 1994. Environmental chemistry of soils. Oxford University Press, New York.

(21) Melville, M.D. and G. Atkinson. 1985. Soil Colour: Its measurement and its designation in models of uniform colour space. J. Soil Sci. 36:495–512.

(22) Moore, D.P. 1971. Physiological effects of pH on roots In Carson, E.W. (ed.) 1974. The plant root and its environment. Proceedings of an institute sponsored by the Southern pRegional Education Board, held at Virginia Polytechnic Institute and State University, July 5–16, 1971, The University Press of Virginia, Charlottesville.

(23) Newman, E.I. 1971. Root and soil water relations In The Plant Root and Its Environment, Proceedings of an institute sponsored by the Southern Regional Education Board, held at Virginia Poltechnic Institue and State University, July 5-16, 1971, E.W. Carson (ed), University Press of Virginia, Charlottesville.

(24) Nielson, K.F. 1971. Roots and Root Temperatures In The plant root and its environment, Proceedings of an institute sponsored by the Southern Regional Education Board, held at Virginia Poltechnic Institute and State University, July 5–16, 1971, edited by E.W. Carson, University Press of Virginia, Charlottesville.

(25) Pierzynski, G.M., J.T. Sims, and G.F. Vance. 1993. Soils and environmental quality, Lewis Publishers, Boca Raton.

(26) Ponte, K.J. and B.J. Carter. 2001. Evaluating differences in soil appearance to aid in developing soil profile photography guidelines. Ph.D. dissertation. Oklahoma State University, Stillwater.

(27) Proceedings international seminar on soil and water utilization, South Dakota State College, Brookings, July 18-August 10, 1962.

(28) Soil Survey Division Staff. 1993. Soil survey manual. USDA, Handbook #18.,U.S. Gov. Prtg. Off., Washington, D.C.

(29) Simonson, R.W. 1993. Soil color standards and terms for field use-history of theirdevelopment. P. 9–13. In J.M. Bigham and E.J. Ciolkosz, (eds) 1993. Soil Color. SSSA Publication #31. SSSA, Madison, WI.

(30) Sparks, D.L. 1995. Environmental Soil Chemistry. Acad. Press, Inc., San Diego.

(31) Stoner, C.H. (ed) 1977. Goodbye to the flush toilet. Rodale Press, Emmaus, PA.

(32) Swanson, P. 2001. Water: the drop of life. Northwood Press, Minnetonka, Minnesota.

www.ingramcontent.com/pod-product-compliance
Lightning Source LLC
Chambersburg PA
CBHW031948190326
41519CB00007B/717